# Cyber-Physical Microgrids

Yan Li

# Cyber-Physical Microgrids

 Springer

Yan Li
Electrical Engineering
Pennsylvania State University
University Park
PA, USA

ISBN 978-3-030-80726-9     ISBN 978-3-030-80724-5    (eBook)
https://doi.org/10.1007/978-3-030-80724-5

This Springer imprint is published by the registered company Springer Nature Switzerland AG
The registered company address is: Gewerbestrasse 11, 6330 Cham, Switzerland

# Preface

Microgrid has been recognized as a promising archetype to improve power system resilience and enhance the reliable operations of low- or medium-voltage power distribution networks. It promotes the high penetration of distributed energy resources into communities to unlock the potentials of microgrids in enhancing the resilience of power grids in a bottom-up way.

In this book, the typical distributed energy resources and communication network will be introduced to build a cyber-physical microgrid system. The modeling of photovoltaic, energy storage units, micro-turbine, wind energy are introduced in detail. To integrate these distributed energy resources into the AC network, power-electronic interfaces are presented as well as their typical control strategies. Typical hierarchical controls for microgrids are summarized, i.e., primary control, secondary control, and tertiary control, where droop control is discussed in detail for island microgrids. Modeling and stability analysis of microgrid systems are also introduced, with a focus on the dynamic modeling and small-signal stability analysis. A typical microgrid test system is designed and provided for demonstrating the dynamic operations of microgrids. The designed system can also be used to test the function of microgrids for remote or isolated places, such as rural communities. The cyber-communication network is also introduced for developing cyber-physical microgrid systems. A general cyber-physical attack scenario is discussed in this book, which includes cyberattacks to microgrid control center, cyberattacks to the cyber-communication network, physical attacks to the microgrids, as well as the propagation of cyber-physical attacks.

The simulation of the designed microgrid system has several applications. For instance, the simulation data can be used for demand response, operation prediction, vulnerability analysis, resilience enhancement, attack detection and mitigation, etc. In additional, novel emerging technologies, such as 5G networking technology, also have potentials to integrate with the physical microgrid to form a new cyber-phyiscal microgrid system, which can be used as a prototype to demonstrate and promote the development of next generation microgrids. Overall, this book can help students and early-stage researchers lay the foundation of cyber-physical microgrids. It also throw light on the understanding and enhancement of the cyber-physical system's resilience, and thus contributing to the sustainable development of the energy field.

University Park, PA, USA                                                                    Yan Li
May 2021

# Contents

1   **Overview of Cyber-Physical Microgrids** ............................... 1
    1.1   Conventional Power Systems ......................................... 1
    1.2   Motivation of Developing Microgrids ............................. 1
    1.3   Physical Layer of CPMs ............................................. 2
    1.4   Cyber Layer of CPMs ................................................ 4
    1.5   Recent Development of Microgrids ............................... 6
    1.6   Scientific Problems of Modern Electric Systems ................. 9
    1.7   Structure of the Book ............................................... 10
    Problems ..................................................................... 11
    References ................................................................... 11

2   **Photovoltaic** ................................................................. 13
    2.1   Introduction of PV .................................................. 13
    2.2   Photovoltaic Physics ................................................ 13
    2.3   Mathematical Model of PV Cell ................................... 15
        2.3.1   Computation of the Photon Current ....................... 16
        2.3.2   Computation of the Shockley
                Temperature-Dependent Diode Current ................... 20
        2.3.3   Computation of the PV Cell Leakage Current .............. 20
    2.4   Mathematical Model of PV Array ................................. 21
    2.5   Output Characteristics of Photovoltaic ........................... 24
    2.6   Maximum Power Point Tracking (MPPT) ....................... 28
        2.6.1   Perturbation and Observation: P&O ...................... 29
        2.6.2   Incremental Conductance ................................. 34
        2.6.3   Varying Step MPPT ....................................... 35
    Problems ..................................................................... 37
    References ................................................................... 39

3   **Energy Storage** ............................................................. 41
    3.1   Battery ............................................................... 41
    3.2   Flywheel .............................................................. 43
    3.3   Supercapacitor ....................................................... 50

3.4    Applications of Energy Storage Units .............................. 51
Problems ................................................................. 52
References ............................................................... 53

**4    Micro-Turbine** ............................................................ 55
4.1    Introduction of Micro-Turbine ...................................... 55
4.2    Synchronous Generator .............................................. 57
       4.2.1   Main Rotating Magnetic Flux .............................. 58
       4.2.2   Armature Reaction ......................................... 60
       4.2.3   Equivalent Circuit of Synchronous Generator .............. 62
4.3    Power Outputs of Micro-Turbine .................................... 65
Problems ................................................................. 73
References ............................................................... 74

**5    Wind Generation** ......................................................... 77
5.1    Introduction of Wind Turbines ...................................... 77
5.2    Fixed-Speed Wind Turbine .......................................... 80
       5.2.1   Mechanical Block .......................................... 81
       5.2.2   Induction Generator ....................................... 82
5.3    Variable-Slip Wind Turbine ......................................... 89
5.4    Doubly-Fed Induction Generator (DFIG) Wind Turbine ........... 91
5.5    Full Converter Wind Turbine ....................................... 94
Problems ................................................................. 94
References ............................................................... 96

**6    Power Electronics Interfaces and Controls** ............................ 99
6.1    Modeling of Inverter ................................................ 99
       6.1.1   Single-Phase Inverter ..................................... 99
       6.1.2   Three-Phase Inverter ..................................... 102
6.2    Controller of Three-Phase Inverter ................................ 103
       6.2.1   Phase Lock Loop (PLL) .................................... 107
       6.2.2   Double Loop Controller ................................... 109
       6.2.3   PWM Generator ............................................ 111
6.3    Chopper ............................................................. 111
       6.3.1   Boost Chopper ............................................ 111
       6.3.2   Buck Chopper ............................................. 114
       6.3.3   Buck-Boost Chopper ....................................... 116
6.4    System Integration of PV ........................................... 120
       6.4.1   Single Stage Integration .................................. 120
       6.4.2   Two-Stage Integration .................................... 122
6.5    System Integration of Energy Storage ............................. 124
       6.5.1   Integration of Battery and Supercapacitor ................ 124
       6.5.2   Integration of Flywheel .................................. 133
6.6    System Integration of Micro-Turbine .............................. 134
6.7    System Integration of Wind Generation ............................ 137
       6.7.1   Equivalent Circuit of Wound Rotor Induction Generator .. 137

|   | 6.7.2 | Equivalent Model in the dq0 Frame | 141 |
|   | 6.7.3 | Vector Control | 144 |
|   | Problems | | 146 |
|   | References | | 148 |

| **7** | **Control of Microgrids** | | 151 |
|   | 7.1 | Microgrids | 151 |
|   | 7.2 | Hierarchical Control | 152 |
|   | 7.3 | Droop Control | 153 |
|   | | 7.3.1 $P - f$ Droop Control | 154 |
|   | | 7.3.2 $Q - V$ Droop Control | 155 |
|   | | 7.3.3 Droop Controls for Multiple DERs | 157 |
|   | 7.4 | $V - f$ Control | 161 |
|   | 7.5 | DER Power Dispatch | 161 |
|   | Problems | | 162 |
|   | References | | 163 |

| **8** | **Modeling and Stability Analysis of Microgrids** | | 165 |
|   | 8.1 | Dynamic Modeling of Microgrids | 165 |
|   | | 8.1.1 $V - f$ Control Time-Domain Simulations | 165 |
|   | | 8.1.2 Droop Control Time-Domain Simulations | 169 |
|   | 8.2 | Small-Signal Stability Analysis of Microgrids | 174 |
|   | | 8.2.1 $V - f$ Control Eigenvalues Calculation | 174 |
|   | | 8.2.2 Droop Control Eigenvalues Calculation | 175 |
|   | Problems | | 175 |
|   | References | | 177 |

| **9** | **Cyber-Communication Network for Microgrids** | | 179 |
|   | 9.1 | Cyber-Communication Network | 179 |
|   | | 9.1.1 Local Area Network | 179 |
|   | | 9.1.2 Metropolitan Area Network | 181 |
|   | | 9.1.3 Wide Area Network | 181 |
|   | | 9.1.4 Wireless Network | 181 |
|   | | 9.1.5 Inter Network | 181 |
|   | 9.2 | Demonstration of Cyber-Physical Microgrids | 182 |
|   | Problems | | 183 |
|   | References | | 184 |

| **10** | **Cyber-Physical Attacks to Microgrids** | | 185 |
|   | 10.1 | Cyber-Physical Attacks | 185 |
|   | 10.2 | Impact of Communication Network | 187 |
|   | 10.3 | Demonstration of Cyber-Attack | 188 |
|   | 10.4 | Demonstration of Physical Attack | 190 |
|   | Problems | | 192 |
|   | References | | 192 |

| **Index** | | | 195 |

# List of Figures

Fig. 1.1    Illustration of conventional power systems ....................... 2
Fig. 1.2    Illustration of cyber-physical microgrids ......................... 3
Fig. 1.3    Modern power system integrated with multiple microgrids ................................................................. 4
Fig. 1.4    A typical microgrid system ......................................... 5
Fig. 1.5    Modern power system integrated with multiple microgrids ................................................................. 6
Fig. 1.6    Modern power system integrated with multiple microgrids ................................................................. 7
Fig. 1.7    Select microgrid assessment and demonstration projects in the U.S. ................................................................. 8
Fig. 1.8    Microgrid penetration capacity in the U.S. in 2019 .............. 9
Fig. 1.9    Scientific problems of modern electric systems with DERs....... 10

Fig. 2.1    Construction of PV array ........................................... 14
Fig. 2.2    Illustration of silicon atoms and electrons ........................ 14
Fig. 2.3    n-type silicon and p-type silicon .................................. 14
Fig. 2.4    Example of electron flows ......................................... 16
Fig. 2.5    Electrical model of PV cell ........................................ 16
Fig. 2.6    The relation of electron's kinetic energy and photon's frequency and light intensity ...................................... 18
Fig. 2.7    The relation of current and photon's frequency and light intensity ............................................................. 18
Fig. 2.8    Illustrations of electrons' energy ................................... 19
Fig. 2.9    Simplified mathematical model of PV cell ....................... 21
Fig. 2.10   Aggregation of PV cells in series ................................. 21
Fig. 2.11   Aggregation of PV cells in parallel .............................. 21
Fig. 2.12   Aggregation of PV cells in series and parallel .................. 22
Fig. 2.13   Simplified aggregation of PV cells in series and parallel ........ 22
Fig. 2.14   Ideal model of PV array ........................................... 24
Fig. 2.15   I-U curves when irradiance changes .............................. 25

Fig. 2.16    P-U curves when irradiance changes ............................. 25
Fig. 2.17    Illustration of the relation of photon current and open
             circuit voltage ........................................................ 26
Fig. 2.18    I-U curves when temperature changes ........................... 28
Fig. 2.19    P-U curves when temperature changes .......................... 28
Fig. 2.20    P&O algorithm ........................................................ 29
Fig. 2.21    Illustration of P&O algorithm .................................... 30
Fig. 2.22    Illustration of P&O under varying environmental
             conditions .............................................................. 32
Fig. 2.23    The first three iterations when the increment of voltage
             is 0.1 V ................................................................ 33
Fig. 2.24    The first three iterations when the increment of voltage
             is 1.0 V ................................................................ 33
Fig. 2.25    The first three iterations when the increment of voltage
             is 8.0 V ................................................................ 34
Fig. 2.26    Incremental conductance method ................................. 35
Fig. 2.27    The iteration steps of varying step MPPT ...................... 36
Fig. 2.28    The changes of step size of varying step MPPT ................ 37
Fig. 2.29    Example of P&O method for Problem 2.8 ...................... 38
Fig. 2.30    Example of P&O method for Problem 2.9 ...................... 39

Fig. 3.1     Short-term discharging model for battery ....................... 42
Fig. 3.2     Generic model for battery .......................................... 43
Fig. 3.3     Typical output of battery voltage ................................. 43
Fig. 3.4     Charge and discharge of flywheel ................................ 45
Fig. 3.5     Constant torque charging for flywheel ........................... 46
Fig. 3.6     Constant power charging for flywheel ........................... 47
Fig. 3.7     Hybrid charging for flywheel ...................................... 49
Fig. 3.8     Linear RC model for supercapacitor .............................. 50
Fig. 3.9     Simplified RC model for supercapacitor ......................... 50
Fig. 3.10    Series RC model for supercapacitor .............................. 51
Fig. 3.11    Three-branch model for supercapacitor .......................... 51
Fig. 3.12    UPS application of energy storage units ......................... 52
Fig. 3.13    Smoothing the power outputs of DERs by using energy
             storage units ........................................................... 52

Fig. 4.1     Single-shaft microturbine ........................................... 56
Fig. 4.2     Double-shaft microturbine .......................................... 57
Fig. 4.3     ABC or ACB phase sequence. (a) Clockwise rotation.
             (b) Counter-clockwise rotation .................................... 58
Fig. 4.4     Magnetic flux linkage. (a) Maximum flux linkage with
             phase A. (b) No flux linkage with phase A ...................... 58
Fig. 4.5     Changes of angle when the flux vector rotates ................. 59
Fig. 4.6     Changes of the magnetic flux and induced voltage over
             time ..................................................................... 60

Fig. 4.7      Single phase equivalent circuit of synchronous generator ........      63
Fig. 4.8      Example of voltage phasors of a synchronous generator .........      63
Fig. 4.9      Simplified voltage phasors of a synchronous generator ..........      65
Fig. 4.10     The relation between $\delta$ and $P$ .....................................      66
Fig. 4.11     Determination of power factors. (**a**) Lagging power
              factor. (**b**) Unity power factor. (**c**) Leading power factor .........      67
Fig. 4.12     Power factors. (**a**) Lagging power factor. (**b**) Unity
              power factor. (**c**) Leading power factor ...........................      68
Fig. 4.13     Test system for Example 2 .........................................      68
Fig. 4.14     The amplitude of the induced voltage $E_A$ is larger than
              $V_A$ .............................................................      72
Fig. 4.15     The amplitude of the induced voltage $E_A$ is equal to $V_A$ .........      73
Fig. 4.16     The amplitude of the induced voltage $E_A$ is smaller than $V_A$ .....      73

Fig. 5.1      Wind power ........................................................      78
Fig. 5.2      Rotor power coefficient $C_p$ .......................................      79
Fig. 5.3      Fixed-speed wind turbine .........................................      80
Fig. 5.4      Components of fixed-speed wind turbine ........................      81
Fig. 5.5      Mechanical block of fixed-speed wind turbine ...................      81
Fig. 5.6      The operation of induction generators ............................      84
Fig. 5.7      Illustration of three magnetic fluxes ..............................      85
Fig. 5.8      Tip speed ratio analysis ............................................      88
Fig. 5.9      Power outputs when the wind speed and rotor's angle
              velocity vary ......................................................      89
Fig. 5.10     Variable-slip wind turbine .........................................      90
Fig. 5.11     Power outputs of the variable-slip wind turbine ...................      90
Fig. 5.12     Doubly-Fed Induction Generator (DFIG) ........................      91
Fig. 5.13     Rotor's speed changes ............................................      92
Fig. 5.14     Rotor's current changes between 12s and 20s ....................      93
Fig. 5.15     Rotor's current changes during $[13.0s, 14.0s]$ ...................      93
Fig. 5.16     Rotor's current changes during $[15.0s, 16.0s]$ ...................      93
Fig. 5.17     Rotor's current changes during $[17.0s, 18.0s]$ ...................      94
Fig. 5.18     Rotor's current changes during $[19.0s, 20.0s]$ ...................      94
Fig. 5.19     Full converter wind turbine .......................................      95

Fig. 6.1      Topology of single-phase inverter ................................      100
Fig. 6.2      Carrier signal and modulation signal for PWM control ..........      100
Fig. 6.3      Relation between modulation index and output RMS
              voltage of the single phase inverter .............................      101
Fig. 6.4      Three-phase inverter ..............................................      102
Fig. 6.5      Relation between modulation index and output RMS
              voltage of the three-phase inverter .............................      103
Fig. 6.6      $dq0$ transformation ..............................................      104
Fig. 6.7      System integration using inverter ................................      108
Fig. 6.8      Example of phase lock loop .......................................      108

| Fig. 6.9 | $dq0$ transformation when the $d$-axis is in phase with the voltage phasor | 108 |
|---|---|---|
| Fig. 6.10 | Example of inner controller for inverter | 111 |
| Fig. 6.11 | Example of boost chopper | 112 |
| Fig. 6.12 | Duty ratio | 112 |
| Fig. 6.13 | Example of buck chopper | 115 |
| Fig. 6.14 | Example of buck-boost chopper | 117 |
| Fig. 6.15 | Topology of single stage PV system integration | 121 |
| Fig. 6.16 | Typical outer controller for PV inverter in the single stage integration | 121 |
| Fig. 6.17 | Topology of two-stage PV system integration | 122 |
| Fig. 6.18 | Example of boost chopper's controller | 123 |
| Fig. 6.19 | Example of two-stage integration for battery and supercapacitor | 124 |
| Fig. 6.20 | Constant current charging control | 125 |
| Fig. 6.21 | Illustration of constant current charging control | 125 |
| Fig. 6.22 | Illustration of constant voltage charging control | 125 |
| Fig. 6.23 | Illustration of combined charging control | 126 |
| Fig. 6.24 | System integration through inverter | 126 |
| Fig. 6.25 | The inverter's output voltage phasor $V_I$ is leading the voltage phasor $V_S$ | 127 |
| Fig. 6.26 | Scenario when inverter's output voltage phasor $V_I$ leads the voltage phasor $V_S$ | 128 |
| Fig. 6.27 | The inverter's output voltage phasor $V_I$ is in phase with the voltage phasor $V_S$ | 128 |
| Fig. 6.28 | Scenarios when the voltage phasor $V_I$ is in phase with $V_S$ | 129 |
| Fig. 6.29 | The inverter's output voltage phasor $V_I$ is lagging the voltage phasor $V_S$ | 129 |
| Fig. 6.30 | Scenarios when the voltage phasor $V_I$ lags the voltage phasor $V_S$ | 129 |
| Fig. 6.31 | Explanation of $V_I \cos(\beta - \alpha) - V_S > 0$ | 130 |
| Fig. 6.32 | Scenarios when $V_I \cos(\beta - \alpha) - V_S > 0$ | 130 |
| Fig. 6.33 | Explanation of $V_I \cos(\beta - \alpha) - V_S = 0$ | 130 |
| Fig. 6.34 | Scenarios when $V_I \cos(\beta - \alpha) - V_S = 0$ | 131 |
| Fig. 6.35 | Explanation of $V_I \cos(\beta - \alpha) - V_S < 0$ | 131 |
| Fig. 6.36 | Scenarios when $V_I \cos(\beta - \alpha) - V_S < 0$ | 132 |
| Fig. 6.37 | Explanation of inverter active and reactive power | 132 |
| Fig. 6.38 | Integration of flywheel system into the AC network | 133 |
| Fig. 6.39 | Controller for the flywheel side converter | 134 |
| Fig. 6.40 | Integration of microturbine system into the AC network | 135 |
| Fig. 6.41 | Active power outputs of generator and inverter | 135 |
| Fig. 6.42 | Three-phase current outputs of inverter | 136 |
| Fig. 6.43 | Capacitor's voltage | 136 |
| Fig. 6.44 | System's frequency | 136 |
| Fig. 6.45 | Stator's and rotor's three-phase magnetic fluxes | 138 |

Fig. 6.46    Stator's electric circuits ............................................... 138
Fig. 6.47    Rotor's electric circuits ............................................... 139
Fig. 6.48    Equivalent circuit to represent $d$-axis calculation ................. 144
Fig. 6.49    Equivalent circuit to represent $q$-axis calculation ................. 144
Fig. 6.50    Total magnetic flux phasor representation ......................... 144
Fig. 6.51    Frames used for vector control ..................................... 146
Fig. 6.52    Inverter's output voltage for Problem 6.9 .......................... 148

Fig. 7.1     Microgrid system ..................................................... 152
Fig. 7.2     Three-layer hierarchical control scheme for microgrids .......... 153
Fig. 7.3     Illustration of $P - f$ droop control ............................... 154
Fig. 7.4     Illustration of $Q - V$ droop control .............................. 156
Fig. 7.5     Battery integration for Example 7.1 ............................... 157
Fig. 7.6     System topology for Example 7.2 ................................. 158
Fig. 7.7     Illustration of $P - f$ droop control for multiple DERs ........... 159
Fig. 7.8     Illustration of $Q - V$ droop control for multiple DERs .......... 160
Fig. 7.9     Example of $V - f$ Control for inverters .......................... 161
Fig. 7.10    System topology for Problem 7.7................................... 163

Fig. 8.1     A typical microgrid test system ................................... 166
Fig. 8.2     Voltage changes of DER buses when the power load
             changes in the Case 1 of $V - f$ control scenario ................. 168
Fig. 8.3     Voltage changes of DER buses when the power load
             changes in the Case 2 of $V - f$ control scenario ................. 169
Fig. 8.4     Voltage changes of DER buses when the power load
             changes in the Case 3 of $V - f$ control scenario ................. 169
Fig. 8.5     Voltage changes of DER buses in the Case 1 of droop
             control scenario .................................................... 170
Fig. 8.6     DER voltage changes in the Case 1 of droop control
             scenario ............................................................ 171
Fig. 8.7     Frequency changes in the Case 1 of droop control
             scenario ............................................................ 171
Fig. 8.8     Voltage changes of DER buses in the Case 2 of droop
             control scenario .................................................... 172
Fig. 8.9     DER voltage changes in the Case 2 of droop control
             scenario ............................................................ 172
Fig. 8.10    Changes of DER active power outputs in the Case 2 of
             droop control scenario .............................................. 173
Fig. 8.11    Changes of DER reactive power outputs in the Case 2 of
             droop control scenario .............................................. 173
Fig. 8.12    Changes of frequency in the Case 2 of droop control
             scenario ............................................................ 173
Fig. 8.13    All eigenvalues of the microgrid test system in the
             $V - f$ control ..................................................... 175

Fig. 8.14    Two pairs of eigenvalues that mainly determine the
             dynamics of the microgrid ......................................... 175
Fig. 8.15    All eigenvalues of the microgrid test system in the droop
             control .............................................................. 176
Fig. 8.16    One pair of eigenvalues that mainly determine the
             dynamics of droop-controlled microgrid ......................... 176
Fig. 8.17    The microgrid system's frequency for Problem 8.4 .............. 176

Fig. 9.1     Illustration of cyber-physical microgrids ........................... 180
Fig. 9.2     Four typical topologies of LAN .................................... 180
Fig. 9.3     Changes of bus voltages in Case 1 ................................. 182
Fig. 9.4     Changes of DER active power in Case 2 ........................... 183
Fig. 9.5     Changes of DER reactive power in Case 2 ......................... 183
Fig. 9.6     Changes of DER bus voltages in Case 2 ........................... 183
Fig. 9.7     Changes of DER bus angles in Case 2 ............................. 184

Fig. 10.1    Generalcyber-physical attack scenario ............................. 186
Fig. 10.2    Voltages under different communication latency .................. 187
Fig. 10.3    Eigenvalues under different communication latency .............. 187
Fig. 10.4    Instantaneous voltage and current responses of Battery
             31 under cyberattack .............................................. 188
Fig. 10.5    Instantaneous voltage and current responses of
             Micro-turbine 20 under cyberattack .............................. 189
Fig. 10.6    Frequency response of the system under cyberattack ............. 189
Fig. 10.7    Micro-turbine 20's voltage and current responses when
             defending the system from cyberattacks .......................... 190
Fig. 10.8    System frequency responses when defending the system
             from cyberattacks ................................................. 190
Fig. 10.9    Instantaneous voltage responses of bus 20 and bus 27
             under physical attack .............................................. 191
Fig. 10.10   RMS values of bus voltage at node 20 and node 27 under
             physical attack .................................................... 191

# List of Tables

Table 2.1    The P&O MPPT method in four different scenarios.............. 31

Table 3.1    Constant torque method and constant power method............. 48

Table 5.1    Synchronous speed ................................................ 83
Table 5.2    Comparison between induction generator and PMSG ............ 87
Table 5.3    Comparison between the fixed speed wind turbine and
             variable-slip wind turbine.......................................... 90

Table 6.1    Changes of duty ratio in the P&O MPPT method ................ 122

Table 8.1    Line impedances between buses in Fig. 8.1 ...................... 167
Table 8.2    Power loads at each bus in Fig. 8.1 ............................. 167
Table 8.3    DER generations at each bus in Fig. 8.1 ........................ 167

# Acronyms

| | |
|---|---|
| AC | Alternating Current |
| CPMs | Cyber-Physical Microgrids |
| DAEs | Differential-Algebraic Equations |
| DC | Direct Current |
| DERs | Distributed Energy Resources |
| DFIG | Doubly-Fed Induction Generator |
| $dq0$ | Direct-quadrature-zero Transformation |
| ICT | Information Communication Technology |
| IGBT | Insulated Gate Bipolar Transistor |
| MCC | Microgrid Coordination Center |
| MPP | Maximum Power Point |
| MPPT | Maximum Power Point Tracking |
| P&O | Perturbation and Observation |
| PCC | Point of Common Coupling |
| PLL | Phase Lock Loop |
| PMSG | Permanent Magnetic Synchronous Generator |
| PV | Photovoltaic |
| PWM | Pulse Width Modulation |
| RMS | Root-Mean-Square |
| RPM | Revolutions Per Minute |
| TSR | Tip Speed Ratio |
| UPS | Uninterruptible Power Supplies |

# Chapter 1
# Overview of Cyber-Physical Microgrids

## 1.1 Conventional Power Systems

Power electricity has become a crucial and indispensable part of our daily life for promoting the growth and development of the society. In the conventional power grids, there are three subsystems [1, 2], namely generation system, transmission system and distribution system, as illustrated in Fig. 1.1. It is a typical centralized unidirectional system. To compare with a power grid integrated with microgrids, we briefly introduce the function of each subsystem in the conventional power grids.

In the generation system, electricity is produced in power plants through burning fossil fuels such as coal, gas and oil, using nuclear material, or utilizing hydroelectric power plants. The function of generation systems is to convert chemical energy to electric energy. It is the source of power electric. Ideally, it can be treated as a specified voltage and frequency source.

Then we convert the energy into a higher voltage level, and transmit the energy through the power lines in the transmission system. Once the energy has travelled through the transmission system, we need to bring the voltage back down. So, the function of transmission system is to deliver electric energy from the generation system to the nearby of customers. Ideally, it can be treated as perfect conductors.

The energy then reaches the final distribution system, where we deliver energy to residential and commercial customers, industry, and other power loads. So, the function of distribution system is to distribute electric energy for customers to use.

## 1.2 Motivation of Developing Microgrids

The conventional top-down monopoly energy configuration is dominated by large-scale fossil and nuclear power plants, which is hard to meet the increasing need of sustainable development of energy, as their emission is detrimental to the environment. Moreover, this centralized configuration which unidirectionally sends

© Springer Nature Switzerland AG 2022                                                                 1
Y. Li, *Cyber-Physical Microgrids*, https://doi.org/10.1007/978-3-030-80724-5_1

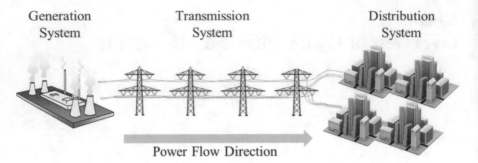

| Generation System | Transmission System | Distribution System |

Power Flow Direction

**Fig. 1.1** Illustration of conventional power systems

electricity to passive consumers is aged and vulnerable to extreme or cascading conditions, e.g., hurricane, earthquake, equipment failure, or cybercrime. For instance, a major blackout hit New York Manhattan on July 13, 2019 [3], which was attributed to a 13 kV cable that burned up at West 64th Street and West End Avenue. This power outage quickly spread out and plunged a broad swath of Manhattan into darkness for up to 5 h on Saturday night, affecting over 73,000 customers and causing tens of millions of dollars' economic loss. The well-known Northeast blackout of 2003 [4], a widespread power outage throughout parts of the Northeastern and Midwestern United States, and the Canadian province of Ontario, affected an estimated 45 million people in eight U.S. states and 10 million people in southern and central Ontario, causing an estimated 10 billion of economic cost. All of these issues deeply block the contribution of power grid to the sustainable development of the whole society.

Since sustainability has become the centre of recent national policies, strategies and development plans of many countries, microgrids have been deployed in recent years to seek an edge toward energy sustainability. Since microgrids highly rely on the coordinated operation and control of its components, an Information Communication Technology (ICT) infrastructure is usually utilized to transmit control signals among its components or between its components and the coordination center, which constitutes **Cyber-Physical Microgrids (CPMs)**. Figure 1.2 illustrates a CPMs system, which includes physical layer, cyber layer, and control center. We will introduce CPMs from both physical layer and cyber layer below.

## 1.3   Physical Layer of CPMs

Distributed Energy Resources (DERs), such as photovoltaic and wind generation, are friendly to our environment and provide more power options for the local customers. An example is given in Fig. 1.3 to show a commercial community has multiple power options. For instance, it could get power electricity from those renewable energy units (e.g., photovoltaic and wind generation), energy storage

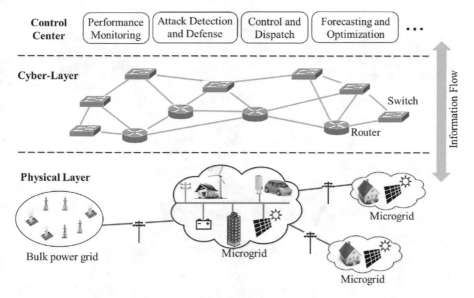

**Fig. 1.2** Illustration of cyber-physical microgrids

units (e.g., battery, super-capacitor, and flywheel), distributed energy units (e.g., micro-turbine and fuel cell), or utility.

To promote the high penetration of DERs, microgrid is usually adopted as an evolving energy landscape [5, 6].

> A microgrid is defined as a group of interconnected loads and DERs within clearly defined electrical boundaries that acts as a single controllable entity with respect to the grid and can connect and disconnect from the grid to enable it to operate in both grid-connected or island mode.

It comprises various DERs, power-electronic interfaces, controllable loads, and monitoring and protection devices. Figure 1.4 shows an example of microgrid, which is composed of photovoltaic, wind generation, energy storage, and loads. When the circuit breaker is open, the microgrid will be operated in the island mode, which could highly reduce the dependency of local customers on utility and improve their local power reliability. While when the circuit breaker is closed, the microgrid will be operated in the grid-connected mode, which could provides ancillary services to the bulk power system, even including the black start of the main grid.

Microgrid is a flexible paradigm to provide affordable, reliable, consistent, and resilient local energy generation and delivery. For one thing, it can help reduce the usage of fossil fuel to protect environment and enable sustainable develop-

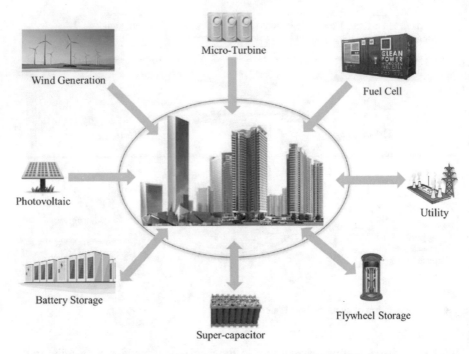

**Fig. 1.3** Modern power system integrated with multiple microgrids

ment. For another thing, microgrids can significantly improve the flexibility and resilience of low- or medium-voltage distribution networks. With the development and penetration of microgrids in the distribution system, the conventional top-down monopoly energy configuration is upgraded to a bi-directional hybrid energy system, as illustrated in Fig. 1.5.

## 1.4  Cyber Layer of CPMs

Communication infrastructure is critical in operating a stable and secure microgrid because microgrid control signals are required to transmit very frequently due to the operation variations, such as those operation changes induced by the intermittence and fluctuation of DERs. As a typical cyber-physical system, the power grid's operation and control highly depends on the acquisition, transmission, and processing of its steady and dynamic behavior. In the bulk power grid, with the rapid development of wide-area measurement systems (WAMS), phasor measurement units (PMU) are widely applied to monitor and control the system through allowing controllers or data analytics to have a global visibility provided by remote signals of PMU. In the fast developing microgrids, for one thing, micro-PMUs are recommended to

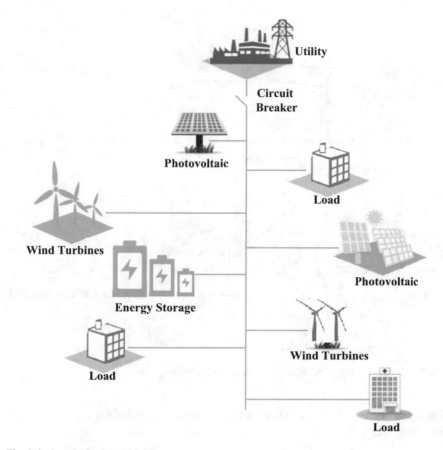

**Fig. 1.4** A typical microgrid system

integrate at key locations to provide systems' measurements for control. For another, information could be directly shared among system player through communication networks.

Actually, several communication networks can be adopted for microgrid communications, such as cellar, WiFi/WiMax, Internet, and power line communications. Figure 1.6 gives an example of the functions of major planes of typical communication networks, i.e., data plane, control plane and application plane. The function of data plane is to enable data transfer to and from microgrid system player. Multiple protocols can be adopted, such as Transmission Control Protocol (TCP) used for data transmission and User Datagram Protocol (UDP) used by programs to send short datagram messages. The function of control plane is The control plane defines the router architecture that is concerned with the network topology. A routing table is usually used in the control plane to determine what to do with the incoming packets from data plane. The application plane includes several microgrid operation

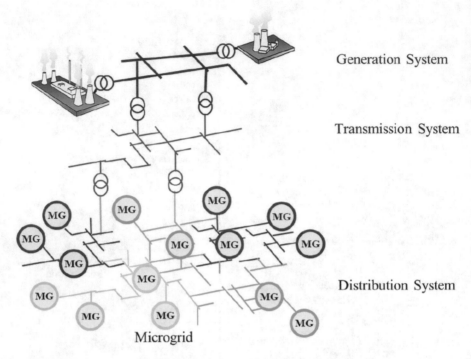

Generation System

Transmission System

Distribution System

Microgrid

**Fig. 1.5** Modern power system integrated with multiple microgrids

functions, such as DER coordination, performance monitoring, and cyberattack detection, etc.

## 1.5   Recent Development of Microgrids

Due to microgrids' benefits to the bulk power systems and local customers as well as their contributions to the sustainable development of the society, microgrids have become increasingly popular and shown their advantages with several policies (e.g. Bills and Acts) passed to support their development. Several policy examples are given below to show U.S. government's support on the development of microgrids and distributed energy resources.

- The Streamlining Energy Efficiency for Schools Act [7] was passed in 2019 to establish programs to improve the energy efficiency of schools. Microgrids will be playing a paramount importance role.
- The European Energy Security and Diversification Act [7] was passed in 2019 to prioritize assistance to develop energy infrastructure in Europe and Eurasia. Assistance includes infrastructure for natural gas, electricity transmission, and

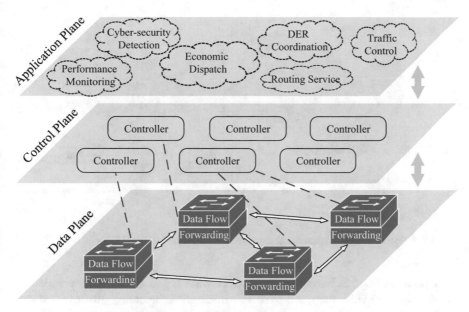

**Fig. 1.6**  Modern power system integrated with multiple microgrids

renewable energy, with the purpose of improving the capacity to transfer gas and electricity within and between regional countries, ensuring regional energy security, and enhancing energy market integration and transparency.

- The North American Energy Security and Infrastructure Act [7] was passed in 2016 to amend the Energy Conservation and Production Act, the Energy Policy and Conservation Act (EPCA), and the Energy Independence and Security Act of 2007 with respect to energy efficiency in buildings and appliances.
- The Solar Technology Roadmap Act [7] was passed in 2009 to conduct a program of research, development, and demonstration for solar technology, including: (1) photovoltaics and related electronic components; (2) solar hot water and solar space heating and cooling; (3) concentrating solar power; (4) lighting systems that integrate sunlight and electrical lighting in complement to each other in common lighting fixtures for the purpose of improving energy efficiency; (5) manufacturability of low cost, high-quality solar energy systems; (6) development of solar technology products that can be easily integrated into new and existing buildings or that are water efficient; and (7) development of storage technologies that can be used to increase the usefulness and value of such technologies.
- The Recovery Act [6] was passed in 2009 to promote the development of new clean energy economy and revitalized infrastructure. U.S. Department of Energy has invested over $ 4.5 billion for modernizing the power grid infrastructure in order to improve its resilience capability.

**Fig. 1.7** Select microgrid assessment and demonstration projects in the U.S.

Several microgrids have been built worldwide based on the policy supports. For instance, in U.S., there are over 160 microgrids, involving all the 50 states. In 2019, over 400 billion kw h power energy in U.S. was generated by DERs of microgrids, which accounted for 17.64% of the domestically produced electricity, and this number is expected to increase to 100% by 2035 or earlier. Figure 1.7 summarizes the microgrid demonstration projects till 2012 [8].

Those microgrid projects are consist of lab- and field-scale R&D test beds, renewable and distributed systems integration (RDSI) projects for peak load reduction, select Smart Grid Demonstration Projects (SGDP), American Recovery and Reinvestment Act of 2009 (ARRA), assessment and demonstration projects jointly supported by the Department of Defense (DoD) and DOE, and projects funded under the DoD Environmental Security Technology Certification Program (ESTCP).

Since 2012, microgrids are being developed and built in the worldwide. For instance, study conducted by Navigant Research also predicts the global microgrid capacity will be growing by 21.4% per annum between 2019 and 2028. In 2016, the U.S. had about 1.6 gigawatts (GW) of installed microgrid capacity. That number increased to 3.2 GW in 2019, and is expected to increase to 4.3 GW by 2020 and 15.8 GW by 2027. Most U.S. microgrid projects are located in Alaska, California, Georgia, Maryland, New York, Oklahoma, and Texas. Figure 1.8 shows the penetration capacity of microgrids in ten U.S. states in 2019 [9].

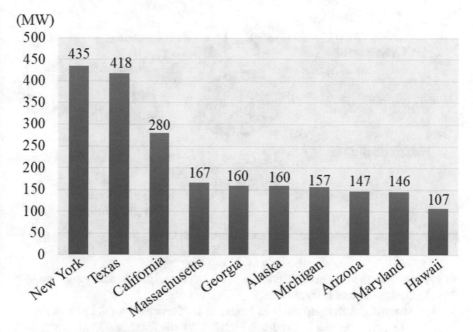

**Fig. 1.8** Microgrid penetration capacity in the U.S. in 2019

## 1.6  Scientific Problems of Modern Electric Systems

Due to the development of CPMs integrated with DERs, the conventional power system's operation and control have been significantly changed. There are several challenging scientific problems that need to pay attention to. For instance, the bidirectional power flow problem caused by the integration of DERs around customers' side and the cyberattack threat induced by the frequent exchange of information between players of microgrids, etc. Figure 1.9 gives ten examples of those critical problems.

A few challenges shown in Fig. 1.9 are discussed below.

1. DER integration. Because most of DERs are connected to system through power electronic device, system inertia is significantly reduced by those interfaces, making microgrid systems very sensitive to disturbances. It is becoming fundamentally important to evaluate systems' stability and other dynamics subject to disturbances.
2. Cyberattacks. Since microgrid operations change frequently, such as operation change from the grid-connected mode to the island mode, correspondingly information needs to be exchanged frequently as well. It gives attacks an opportunity to manipulate the system through touching the communication network used for exchanging information. So, how to effectively defend against

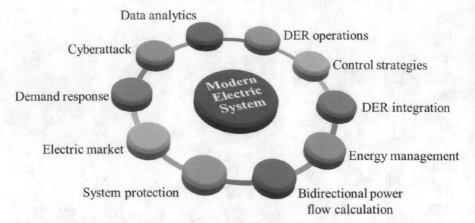

**Fig. 1.9** Scientific problems of modern electric systems with DERs

cyberattacks on the communication network or protect data security and integrity is becoming another critical issue.

3. Data analytics. Although, analysis based on mathematical model is accurate, it is hard or even impossible to build a high fidelity mathematical model in reality. An inspiring question is how could we leverage machine learning or artificial intelligence to build a data-driven cyber-physical microgrid model and then use it to provide operation commands for system operators.

## 1.7   Structure of the Book

This book introduces the modeling, control and operation of cyber-physical micro-grids. Its chapters are organized as below.

- Chapter 1 presents an overview of the cyber-physical microgrids system, includ-ing distributed energy resources, physical layer, cyber layer, recent development, and scientific problems.
- Chapter 2 introduces the modeling of photovoltaic. The Maximum Power Point Tracking (MPPT) control strategies are also introduced in detail, with a focus on the perturbation and observation method, incremental conductance method, and an adaptive MPPT method.
- Chapter 3 introduces the modeling of energy storage units, including battery, flywheel, and supercapacitor. Typical applications of energy storage units are given as examples in the end.
- Chapter 4 introduces the modeling of micro-turbine system with a focus on the single-shaft microturbine. Synchronous generator is also introduced to convert the mechanical energy to electric energy.

- Chapter 5 introduces the modeling of wind generation. Four typical types of wind turbines are presented, including fixed-speed wind turbine, variable-slip wind turbine, doubly-fed induction generator (DFIG) turbine, and full converter wind turbine.
- Chapter 6 introduces power electronics interfaces used to integrate DERs into microgrids. The modeling of inverters and choppers are introduced as well as the typical double loop control strategy.
- Chapter 7 introduces the typical hierarchical control strategy. Droop control is presented to control island microgrids. VF control strategy is introduced to control those dispatchable units.
- Chapter 8 presents the modeling of microgrids including dynamic modeling and small-signal stability analysis. Differential algebraic equations are built to calculate and analyze microgrid dynamics under disturbances.
- Chapter 9 presents the cyber-communication network for microgrids. The construction of communication network is introduced for building a cyber-physical microgrid system.
- Chapter 10 summarizes common cyberattacks in microgrids and discusses the impacts of cyber-physical attacks on the dynamic operations of microgrids.

## Problems

**1.1** Describe the function of three subsystems of the conventional power grids.

**1.2** Why is it necessary to develop microgrid systems?

**1.3** Describe a microgrid system and give an example.

**1.4** What is the function of the cyber-layer of CPMs?

**1.5** What is the typical control strategy of the PV systems?

**1.6** Describe the bi-directional power flow phenomenon and explain what causes it.

## References

1. Machowski, J., Lubosny, Z., Bialek, J. W., & Bumby, J. R. (2020). *Power system dynamics: stability and control*. John Wiley & Sons.
2. Kundur, P., Balu, N. J., & Lauby, M. G. (1994). *Power system stability and control* (Vol. 7). McGraw-Hill.
3. https://www.pulse.ng/the-new-york-times/world/power-failure-hits-manhattans-west-side-leaving-62000-customers-in-the-dark/nsrk4n4
4. https://web.archive.org/web/20160518020430/http://www.prnewswire.com/news-releases/what-caused-the-power-blackout-to-spread-so-widely-and-so-fast-genscapes-unique-data-will-help-answer-that-question-70952022.html

5. Bossart, S. (2012). DOE perspective on microgrids. In *Advanced Microgrid Concepts and Technologies Workshop*.
6. Anadon, L. D., GaLLaGher, K. S., & Bunn, M. G. (2009). DOE FY 2010 budget request and Recovery Act funding for energy research, development, demonstration, and deployment: Analysis and recommendations. Cambridge: Report for Energy Technology Innovation Policy research group, Belfer Center for Science and International Affairs, Harvard Kennedy School.
7. https://www.congress.gov/
8. Ton, D. T., & Smith, M. A. (2012). The US department of energy's microgrid initiative. *The Electricity Journal, 25*(8), 84–94.
9. https://www.statista.com/statistics/1100458/-capacity-of-us-microgrids-by-state/

# Chapter 2
# Photovoltaic

## 2.1 Introduction of PV

PV is a direct means to convert sunlight to electrical energy [1–3]. The typical PV conversion efficiency is in the range of 10%–20% [4, 5]. Because PV generates DC power instead of AC power [6], power-electronic interface can be used to converter DC output to AC power [7–10], then PV energy can be integrated into the power system. There are multiple types of PV, such as ground-mounted PV panel [11], rooftop mounted PV array [12], wall mounted [13] or floating [14], etc. The fundamental operation of different PV panels are very similar to each other, as described below.

Solar cells make up PV modules. PV array consists of several PV modules in series and parallel connections, as shown in Fig. 2.1. The most common solar cells are made from silicon [15], which is a semiconductor that is the second most abundant elements on Earth. Crystalline silicon is the dominant semiconducting material used for the production of solar cells [16]. Crystalline silicon is sandwiched between conductive layers.

## 2.2 Photovoltaic Physics

Photovoltaics converts light into electricity at the *atomic* level [17]. Silicon exhibits a property known as the **Photoelectric Effect** that causes them to absorb photons of light and release electrons [18]. When these free electrons are captured, electric current is generated.

Generally, each silicon atom is connected to its neighbors by four strong bonds, as illustrated in Fig. 2.2. Without sunlight, the electrons are kept in place, i.e., no electron can flow. That is no current will be generated when we build a closed-loop circuit. So the essential idea to generate current is to have electrons move. To realize that goal, we need to look at the silicon cell in more detail.

© Springer Nature Switzerland AG 2022
Y. Li, *Cyber-Physical Microgrids*, https://doi.org/10.1007/978-3-030-80724-5_2

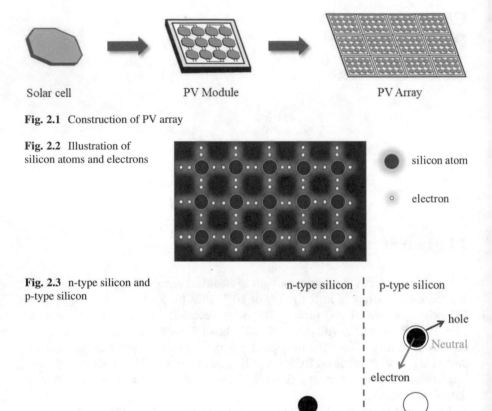

Solar cell                    PV Module                                   PV Array

**Fig. 2.1**  Construction of PV array

**Fig. 2.2**  Illustration of
silicon atoms and electrons

**Fig. 2.3**  n-type silicon and
p-type silicon

A silicon cell uses two different layers of silicon [19], i.e., n-type silicon and p-type silicon, as illustrated in Fig. 2.3. An n-type silicon has extra electrons with negative charge. A p-type silicon has extra spaces for electrons, which is called holes with positive charge. The junction of the two types of silicon is called P/N junction. Electrons can wander across the P/N junction. When electrons are wandering, a positive charge is left on one side (p-side), creating a negative charge on the other (n-side). But in this scenario, even electrons can wander, we still do not have current when we connect the cell to build a closed-loop circuit, as generating current needs tons of electrons to regularly flow from one terminal to the other. The key to generate current is to have more free electrons. That means we need energy to help electrons move!

The energy we talk about here is sunlight. The light is made of photons [20]. In physics, a photon is a bundle of electromagnetic energy. Photons are not made up of smaller particles. They are the basic unit of nature called an elementary particle. The photon is sometimes referred to as a "quantum" of electromagnetic energy, such as light and radio waves. Photons are always in motion and they move at the speed of light in vacuum, 299,792,458 m/s. Photons have several important features. The features relative to PV are summarized below [21].

- Photons are massless.
- They have no electric charge.
- The energy and momentum that carried by photons are dependent on their frequency.
- They can have interactions with other particles such as electrons, such as the Compton effect or Compton scattering [22, 23].
- Compton scattering, discovered by Arthur Holly Compton, is the scattering of a photon by a charged particle, usually an electron. If it results in a decrease in energy of the photon, it is called the Compton effect.

Based on Compton effect, we can see that when one photon strikes the silicon cell with enough energy, it can knock an electron from its bond. A hole will be left. The negative charged electron and location of the positively charged hold are free to move around. But because of the electric field at the P/N junction, they will only go one way, i.e., the electron is drawn to the n-side while the hole is drawn to the p-side. The mobile electrons are collected by thin metal fingers at the top of the cell. If we have an external circuit, then we could have current, as the example given in Fig. 2.4.

## 2.3   Mathematical Model of PV Cell

As introduced, a PV array is eventually made up by several solar cells. To properly model a PV array, we need to first model a solar cell. A solar cell can be modeled in different ways based on the study and research perspectives. For studying the dynamic operations, an electrical model of PV cell is given in Fig. 2.5.

The parameters involved in Fig. 2.5 are explained below.

- $I_{ph}$ is the photon current, which depends on the light intensity and its wavelength.
- $I_d$ is the Shockley temperature-dependent diode current [24].
- $I_p$ is the PV cell leakage current.
- $I$ and $U$ are output current and voltage.

The computation of $I_{ph}$, $I_d$, and $I_p$ will be introduced in the following subsections. By using KCL, we can get the following calculation of the output current $I$.

$$I = I_{ph} - I_d - I_p. \tag{2.1}$$

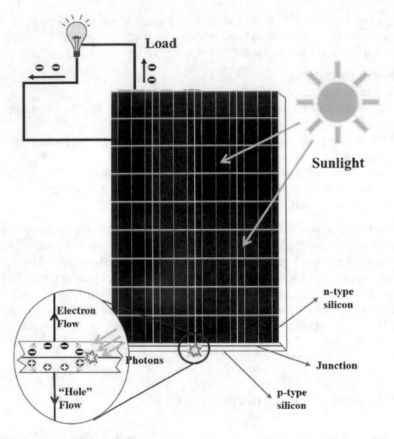

**Fig. 2.4** Example of electron flows

**Fig. 2.5** Electrical model of
PV cell

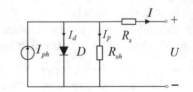

## 2.3.1   Computation of the Photon Current

The photon current is the power source of PV array. This current is determined by
the light intensity and its wavelength. Mathematically, the photon current can be
calculated by using (2.2).

$$I_{ph} = \frac{S}{S_{ref}}\Big(I_{phref} + C_T(T - T_{ref})\Big). \tag{2.2}$$

where,

- $S$ is the irradiance value.
- $S_{ref}$ is the reference value of irradiance, typically $1000\,\text{W/m}^2$ (watts per square meter).
- $I_{phref}$ is the photon current under the reference irradiance.
- $C_T$ is the temperature coefficient [25].
- $T$ is the temperature in Kelvin scale.
- $T_{ref}$ is reference value of temperature, typically $298\,\text{K}$ which is equal to $24.85\,^\circ\text{C}$ (Celsius) or $76.73\,^\circ\text{F}$ (Fahrenheit)).

### 2.3.1.1 Irradiance's Impact

Assume the temperature is constant, from (2.2), we can see when the irradiance increases, the photon current $I_{ph}$ will increase. Why? We will explain the reason from the following two aspects.

*Mathematical Analysis* From the mathematical analysis, it is straightforward that the increase of irradiance will cause the photon current to increase.

*Physics Analysis* The generation of the photon current depends on the light intensity and its wavelength $\lambda$. To explain this statement, we need to look from the energy perspective. Considering the *Conservation of Energy* that is the total energy of an isolated system remains constant, we can express the photon's energy by using (2.3).

$$E_{photon} = E_{electron} + W, \tag{2.3}$$

where,

- $E_{photon}$ is the energy of photon.
- $E_{electron}$ is the kinetic energy of electron.
- $W$ is the energy used to overcome the binding energy of the electron in silicon.

The photon energy $E_{photon}$ can be calculated via (2.4), where $h$ is Planck's constant and $f$ is the wave frequency. The kinetic energy of electron $E_{electron}$ can be calculated via (2.5), where $m$ is the mass of electron and $v$ is the speed.

$$E_{photon} = hf = h\frac{c}{\lambda}. \tag{2.4}$$

$$E_{electron} - \frac{1}{2}mv^2. \tag{2.5}$$

The relation of the kinetic energy of electron and photon's frequency and light intensity is usually depicted by Fig. 2.6, from which we can see that:

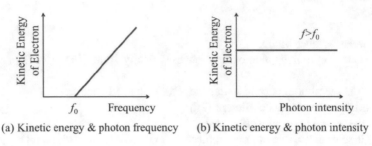

(a) Kinetic energy & photon frequency    (b) Kinetic energy & photon intensity

**Fig. 2.6** The relation of electron's kinetic energy and photon's frequency and light intensity

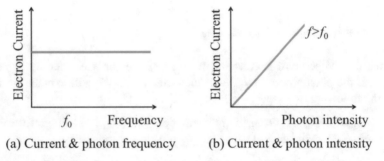

(a) Current & photon frequency    (b) Current & photon intensity

**Fig. 2.7** The relation of current and photon's frequency and light intensity

- The kinetic energy of the electrons is linearly proportional to the frequency of the light radiation above a threshold value. This conclusion is easily to justify. We substitute (2.4) into (2.3), then we can get $E_{electron} = E_{photon} - W = hf - W$, which shows the linear proportional relation.
- There is a frequency threshold below which we have no free electrons, i.e., no current will be generated. For silicon, the highest wavelength when the energy of photon is still big enough to produce free electrons is $1.15\,\mu m$.
- The kinetic energy is independent of the intensity of the radiation, as the intensity of the light means the number of photons.

The relation of the current and photon's frequency and light intensity can be depicted by Fig. 2.7, from which we can see that:

- The electric current is independent of the frequency of the radiation above the threshold value.
- Electric current is proportional to the number of photons (i.e. the intensity of the light) above the threshold value.
- It further verifies the photon current $I_{ph}$ depends on the light intensity and its wavelength $\lambda$, namely frequency.

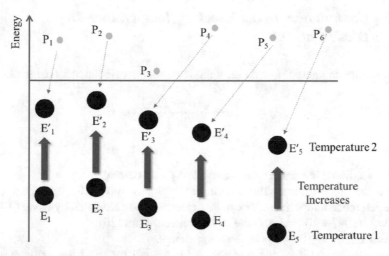

**Fig. 2.8** Illustrations of electrons' energy

### 2.3.1.2  Temperature's Impact

Assume the irradiance is constant, from (2.2), we can see when the temperature increases, the photon current $I_{ph}$ will also increase. Why? We can explain it from the following two aspects.

*Mathematical Analysis*  From the mathematical analysis, it is straight forward that the increase of temperature will cause the photon current to increase.

*Physics Analysis*  The temperature's increase will also increase the energy of photon. The explanation is illustrated in Fig. 2.8.

At Temperature 1, we assume the electrons' energy are expressed by $E$, e.g., $E_1$ and $E_2$. In this scenario, the photons' energy is not enough to free those electrons. When the temperature increases to the level of 2, the electrons' energy will also correspondingly increase, say $E'$ level as shown in Fig. 2.8. In this scenario, the photons' energy is able to free some of those electrons, e.g., the left two electrons. Therefore, current can be produced when the PV connects to a closed-loop electric circuit. Hence, as the temperature increases, the photon current $I_{ph}$ will also increase.

*Note 2.1*  The value of $C_T$ in (2.2) is very small, so the photon current is mainly determined by the irradiance value.

### 2.3.2  Computation of the Shockley Temperature-Dependent Diode Current

The Shockley temperature-dependent diode current is calculated by using (2.6).

$$I_d = I_S \Big( \exp \frac{q(U + I R_S)}{\eta k T} - 1 \Big), \tag{2.6}$$

where,

- $I_d$ is the Shockley temperature-dependent diode current.
- $I_S$ is the diode reverse saturated current, typically $100 \, \text{pA}$ for the silicon cell at the reference temperature. $I_S$ can be expressed and calculated by using (2.7).
- $k = 1.38047 * 10^{-23}$ J/K is the Boltzmann constant [26].
- $q = 1.60201 * 10^{-19}$ C is the electron charge.
- $\eta$ is an empirical ideal constant adjusted to around 1.2 and 1.8 for silicon.
- $\frac{kT}{q} = V_T$ is the temperature equivalent parameter.
- $E_g$ is the band gap. 1.14 for silicon.

$$I_S = I_{Sref} \Big( \frac{T}{T_{ref}} \Big)^3 \exp \Big\{ \frac{q E_g}{\eta k} \Big( \frac{1}{T_{ref}} - \frac{1}{T} \Big) \Big\}. \tag{2.7}$$

From (2.6) and (2.7), we can see that as the temperature increases, $I_d$ will significantly increase.

### 2.3.3  Computation of the PV Cell Leakage Current

In Fig. 2.5, by using KVL, the PV cell's leakage current can be got as follows.

$$I_p = \frac{U + I R_S}{R_{sh}}, \tag{2.8}$$

where,

- $U$ is the output voltage of the PV cell.
- $I$ is the output current.
- $R_S$ is the impedance in series.
- $R_{sh}$ is the parallel impedance.

Note 2.2 When we ignore the leakage current and the series impedance $R_S$, the PV cell's current output can be simplified, as shown in (2.9). And the simplified mathematical model for the PV cell is given in Fig. 2.9.

**Fig. 2.9** Simplified mathematical model of PV cell

**Fig. 2.10** Aggregation of PV cells in series

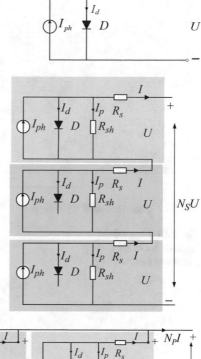

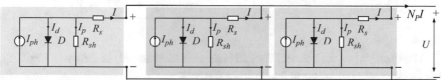

**Fig. 2.11** Aggregation of PV cells in parallel

$$I = I_{ph} - I_d$$

$$= \frac{S}{S_{ref}}\left(I_{phref} + C_T(T - T_{ref})\right) - I_S\left(\exp\left(\frac{qU}{\eta kT}\right) - 1\right). \tag{2.9}$$

## 2.4 Mathematical Model of PV Array

A PV array is made up by several PV cells. So, when we connect PV cells in series, as shown in Fig. 2.10, we can increase the output voltage without changing its current output, where $N_S$ is the number of PV cell in series.

When we connect PV cells in parallel, as shown in Fig. 2.11, we can increase the output current without changing its voltage output, where $N_P$ is the number of PV cell in parallel.

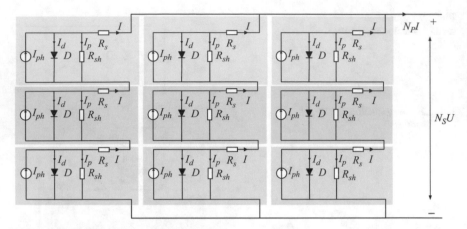

**Fig. 2.12** Aggregation of PV cells in series and parallel

**Fig. 2.13** Simplified aggregation of PV cells in series and parallel

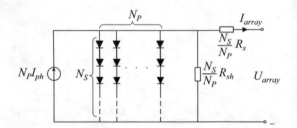

Therefore, when we connect PV cells in series and parallel, as shown in Fig. 2.12, we can increase both current and voltage outputs.

Since the model given in Fig. 2.12 is complicated for computation and analysis, we can further simplify the model to Fig. 2.13. Correspondingly, the PV array's current output can be expressed by referring to the PV cell's current output.

The PV array's current output is computed in (2.10).

$$I_{array} = \hat{I}_{ph} - \hat{I}_d - \hat{I}_p, \tag{2.10}$$

where,

- $\hat{I}_{ph}$ is the total photon current, i.e., $\hat{I}_{ph} = N_P I_{ph}$
- $\hat{I}_d$ is the total Shockley temperature-dependent diode current, which can be calculated by $N_P I_d$.
- $\hat{I}_p$ is the total leakage current, i.e., $N_P I_p$.

The calculations of $N_P I_p$ and $N_P I_d$ are derived below. Based on (2.8), $N_P I_p$ can be calculated through (2.11).

$$\hat{I}_p = N_P I_p = N_P \frac{U + I R_S}{R_{sh}}. \tag{2.11}$$

Then, the question becomes how to use $U_{array}$ and $I_{array}$ to represent $U$ and $I$. Apply KVL to Fig. 2.13, we can get the voltage over the impedance $\frac{N_S}{N_P} R_{sh}$ is:

$$U_{\frac{N_S}{N_P} R_{sh}} = U_{array} + I_{array} \frac{N_S}{N_P} R_S. \tag{2.12}$$

Thus, the voltage over one $R_{sh}$ should be equal to $\frac{1}{N_S} U_{\frac{N_S}{N_P} R_{sh}}$, i.e.,

$$U + I R_S = \frac{1}{N_S} U_{\frac{N_S}{N_P} R_{sh}} = \frac{U_{array}}{N_S} + \frac{I_{array}}{N_P} R_S. \tag{2.13}$$

Substitute (2.13) to (2.11), we can get:

$$\hat{I}_p = N_P I_p = \frac{N_P}{R_{sh}} \left( \frac{U_{array}}{N_S} + \frac{I_{array}}{N_P} R_S \right). \tag{2.14}$$

Based on (2.6), $N_P I_d$ can be calculated through (2.15).

$$\hat{I}_d = N_P I_d = N_P I_S \left( \exp \frac{q(U + I R_S)}{\eta k T} - 1 \right). \tag{2.15}$$

Considering (2.13), $N_P I_d$ can be updated as,

$$\hat{I}_d = N_P I_d = N_P I_S \left( \exp \left\{ \frac{q}{\eta k T} \left( \frac{U_{array}}{N_S} + \frac{I_{array}}{N_P} R_S \right) \right\} - 1 \right). \tag{2.16}$$

Therefore, based on (2.14) and (2.16), the PV array's current output can be rewritten as,

$$
\begin{aligned}
I_{array} &= \hat{I}_{ph} - \hat{I}_d - \hat{I}_p \\
&= N_P I_{ph} - N_P I_S \left( \exp \left\{ \frac{q}{\eta k T} \left( \frac{U_{array}}{N_S} + \frac{I_{array}}{N_P} R_S \right) \right\} - 1 \right) \\
&\quad - \frac{N_P}{R_{sh}} \left( \frac{U_{array}}{N_S} + \frac{I_{array}}{N_P} R_S \right).
\end{aligned}
\tag{2.17}
$$

When the leakage current and the series impedance $R_S$ are ignored, the PV array's current output can be calculated as (2.18). And an ideal mathematical model for the PV array is given in Fig. 2.14.

$$I_{array} = \hat{I}_{ph} - \hat{I}_d = N_P I_{ph} - N_P I_S \left( \exp \left( \frac{q U_{array}}{\eta k T N_S} \right) - 1 \right). \tag{2.18}$$

**Fig. 2.14** Ideal model of PV array

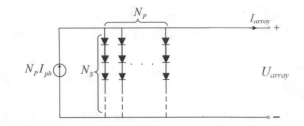

## 2.5  Output Characteristics of Photovoltaic

The output characteristics of photovoltaic will be introduced by using the following two examples.

**Example 2.1**
As shown in Fig. 2.14, an ideal PV array has 20 PV cell connected in parallel and 40 PV cell in series. The photon current under the reference condition 1000 W/m$^2$ and 298 K is 3.35 A. $\eta$ is 1.5, $k$ is $1.38047 * 10^{-23}$ J/K, $q$ is $1.60201 * 10^{-19}$ C. $C_T$ is $2.1775 * 10^{-5}$ A/K, $E_g$ is 1.14, and $I_{sref}$ is 100 pA, i.e., $100 * 10^{-12}$ A.
Questions:
Question 1: What are the voltage, current, and active power outputs when the irradiance is 900 W/m$^2$ and the temperature is 298 K?
Question 2: What are the voltage, current, and active power outputs when the irradiance is 1000 W/m$^2$ and the temperature is 298 K?
Question 3: What are the voltage, current, and active power outputs when the irradiance is 1100 W/m$^2$ and the temperature is 298 K?

Solution: In order to calculate the voltage, current, and active power, we can see (2.18) is the essential equation. When we look at (2.18) in more detail, we can see if we actively change the output voltage $U_{array}$, we can correspondingly get the output current $I_{array}$. So, when the voltage changes from 0, which is the short circuit, to the open circuit value, the current output can be obtained. Figure 2.15 demonstrates the current values as we change $U_{array}$ in (2.18). We call those curves as 'I-U' curves. Three irradiance levels are provided as examples, namely, 900 W/m$^2$, 1000 W/m$^2$, and 1100 W/m$^2$.

When we get the current and voltage, it is easy to compute the active power output, i.e.,

$$P_{array} = I_{array} * U_{array}. \tag{2.19}$$

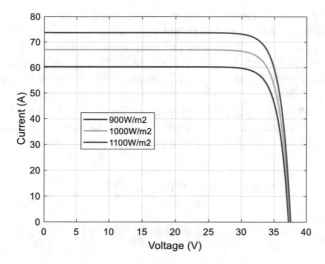

**Fig. 2.15**  I-U curves when irradiance changes

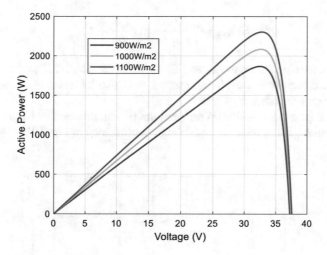

**Fig. 2.16**  P-U curves when irradiance changes

Then, we can correspondingly generate the active power output curves, as shown in Fig. 2.16. We call those curves as 'P-U' curves.

From Figs. 2.15 and 2.16, we can see that:

- As the irradiance increases, the short circuit current increases a lot.
- As the irradiance increases, the open circuit voltage increases, but does not increase a lot.
- As the irradiance increases, the power output of PV array increases.

The first finding is explained below. From Fig. 2.14, we can see that when we short circuit the PV array, the output voltage $U_{array}$ should be equal to zero. Substitute $U_{array} = 0$ to (2.18), we can see the output current $I_{array}$ should be the photon current $N_P I_{ph}$, where $I_{ph}$ is given in (2.2). From (2.2), when the irradiance increases, the photon current will also increase; and thus, the short circuit current will increase as well.

The second finding is explained below. From Fig. 2.14, we can see that when the PV array is open circuit, the output current $I_{array}$ should be equal to zero. Substitute $I_{array} = 0$ to (2.18), the following two equations can be obtained.

$$N_P I_{ph} - N_P I_S \left( \exp\left(\frac{q U_{array}}{\eta k T N_S}\right) - 1 \right) = 0, \tag{2.20}$$

$$I_{ph} = I_S \left( \exp\left(\frac{q U_{array}}{\eta k T N_S}\right) - 1 \right). \tag{2.21}$$

Then, we can rewrite (2.21) to get the expression of $U_{array}$:

$$U_{array} = \frac{\eta k T N_S}{q} \ln\left(\frac{I_{ph}}{I_S} + 1\right). \tag{2.22}$$

Considering the calculation of $I_{ph}$ and $I_S$ shown in (2.2) and (2.7), we can see when the irradiance increases, the photon current will also increase; and thus, the open circuit voltage will increase as well.

When the irradiance is 1000 W/m$^2$ and the temperature is 298 K, the relation of photon current $I_{ph}$ and open circuit voltage is illustrated in Fig. 2.17.

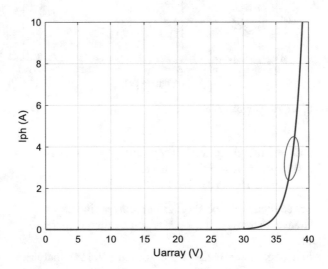

**Fig. 2.17** Illustration of the relation of photon current and open circuit voltage

From Fig. 2.17, we can see that when the irradiance increase causes the photon current $I_{ph}$ to significantly increase, the open circuit voltage $U_{array}$ does not increase a lot, as the circled section in Fig. 2.17. The circled section is corresponding to the open circuit voltage range [35, 40] as shown in Figs. 2.15 and 2.16. ∎

**Example 2.2**
As shown in Fig. 2.14, an ideal PV array has 20 PV cell connected in parallel and 40 PV cell in series. The photon current under the reference condition $1000\,W/m^2$ and 298 K is 3.35 A. $\eta$ is 1.5, $k$ is $1.38047 * 10^{-23}\,J/K$, $q$ is $1.60201 * 10^{-19}\,C$. $C_T$ is $2.1775 * 10^{-5}\,A/K$, $E_g$ is 1.14, and $I_{sref}$ is 100 pA, i.e., $100 * 10^{-12}\,A$.
Questions:
Question 1: What are the voltage, power and current outputs when the temperature is 250 K and the irradiance is $1000\,W/m^2$?
Question 2: What are the voltage, power and current outputs when the temperature is 298 K and the irradiance is $1000\,W/m^2$?
Question 3: What are the voltage, power and current outputs when the temperature is 350 K and the irradiance is $1000\,W/m^2$?

Solution: Similar to Example 2.1, when we actively change the output voltage $U_{array}$ in (2.18), we can correspondingly get the output current $I_{array}$. So, when the voltage changes from 0, which is the short circuit, to the open circuit value, the current output can be obtained. Figure 2.18 demonstrates the current values as we change $U_{array}$ in (2.18). Three temperature values are provided as examples, namely, 250, 298, and 350 K. After we get voltage and current, the 'P-U' curves can also be obtained, as shown in Fig. 2.19.
From Figs. 2.18 and 2.19, we can see that:

- As the temperature increases, the open circuit voltage decreases.
- As the temperature increases, the short circuit current does not change much.
- As the temperature increases, the power output of PV array decreases.

The first finding is explained below. Considering the calculation of $I_{ph}$ and $I_S$ shown in (2.2) and (2.7), we can see when the temperature increases, the photon current $I_{ph}$ will increase as well as the diode reverse saturated current $I_S$. Further studies show $I_S$ will significantly increase as the temperature increases. Hence, according to (2.22), the open circuit voltage $U_{array}$ decreases.
The second finding is explained below. When we short circuit the PV array, $U_{array} = 0$, then the output current $I_{array}$ should be the photon current $N_P I_{ph}$, where $I_{ph}$ is given in (2.2). From (2.2), when the temperature increases, the photon current will increase; but the temperature coefficient $C_T$ is usually very small (e.g., $2.1775 * 10^{-5}\,A/K$ in this example), eventually, the short circuit current does not change much as the temperature changes. ∎

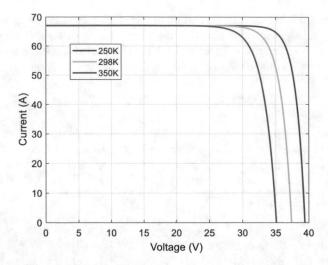

**Fig. 2.18** I-U curves when temperature changes

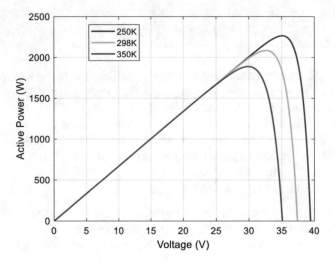

**Fig. 2.19** P-U curves when temperature changes

## 2.6 Maximum Power Point Tracking (MPPT)

From the power output curves of PV as shown in Figs. 2.16 and 2.19, we can see it is necessary to operate the PV array at the maximum power point (MPP) under varying environmental conditions. There are several Maximum Power Point Tracking (MPPT) methods [27–30], e.g., Perturbation and Observation–P&O, Incremental Conductance, Ripple Correlation Control, Current Sweep, Fuzzy methods, and AI method. Two methods, P&O and Incremental Conductance, are introduced here, with an adaptive MPPT method provided as a detailed example [28].

### 2.6.1   *Perturbation and Observation: P&O*

Perturbation and Observation (P&O) method is also called Hill Climbing technique. It introduces a perturbation in the operating voltage of the PV array.

- If there is an increase in the PV active power output after the voltage perturbation, the subsequent perturbation should be kept the same to reach the MPP.
- If there is a decrease in the PV active power output after the voltage perturbation, the subsequent perturbation should be reversed.

A P&O algorithm is introduced in Fig. 2.20. To implement the P&O algorithm, we need to compare the current step's PV active power output and voltage with the previous step, where four calculation paths are involved.

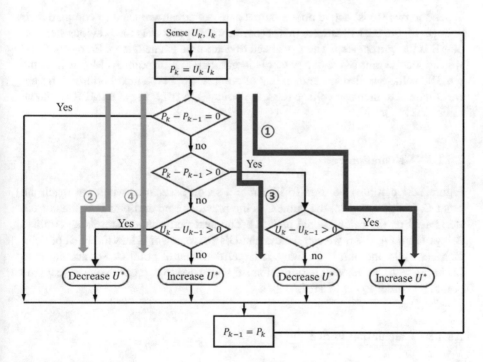

**Fig. 2.20**  P&O algorithm

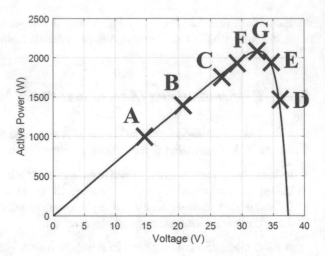

**Fig. 2.21** Illustration of P&O algorithm

### 2.6.1.1   Calculation Path 1

When the two steps' active power outputs are not equal, we need to compare their voltage. Figure 2.21 illustrates this scenario, where point A is the previous step and pint B is the current step. Then, we need to determine the next step. From Fig. 2.21, we can see the pint B's active power is larger than that of point A. Meanwhile, the pint B's voltage is also larger than that of point A. So, in the next perturbation step, we still need to increase voltage (e.g., the point C) in order to get to MPP, as shown in Fig. 2.21 ①.

### 2.6.1.2   Calculation Path 2

Figure 2.21 can also be used to illustrate a second scenario, when we reach the point C. In this scenario, the point C is the previous step and pint D is the current step based on the calculational path ①. The next perturbation step is determined below. From Fig. 2.21, we can see the pint D's active power is less than that of point C. Meanwhile, the pint D's voltage is larger than that of point C. So, according to ② in Fig. 2.20, in the next perturbation step, we need to decrease voltage (e.g., the point E) in order to get to MPP.

### 2.6.1.3   Calculation Path 3

When we compare the points D (previous step) and E (current step), we can see the pint E's active power is larger than that of point D, but its voltage is smaller than point D. In this scenario, according to ③ in Fig. 2.20, we need to decrease voltage (e.g., the point F) in order to get to MPP.

**Table 2.1** The P&O MPPT method in four different scenarios

|   | Voltage perturbation | Active power change | Next voltage perturbation |
|---|---|---|---|
| ① | Positive | Positive | Positive |
| ② | Positive | Negative | Negative |
| ③ | Negative | Positive | Negative |
| ④ | Negative | Negative | Positive |

### 2.6.1.4   Calculation Path 4

When we compare the points E (previous step) and F (current step), we can see the pint F's active power is smaller than that of point E, and its voltage is also smaller than point E. In this scenario, according to ④ in Fig. 2.20, we need to increase voltage (e.g., the point F) in order to get to MPP, i.e., the point G.

*Note 2.3* From the analysis of the above four calculation paths, we can summarize the next voltage perturbations under different scenarios, as given in Table 2.1.

From Table 2.1, we can see the direction (positive or negative) of the next voltage perturbation can be determined by using the following (2.23).

$$D_{\Delta U}^{i+1} = D_{\Delta P}^{i} * D_{\Delta U}^{i}. \tag{2.23}$$

where,

- $D_{\Delta U}^{i+1}$ is the direction of the next voltage perturbation.
- $D_{\Delta P}^{i}$ is the direction of the active power.
- $D_{\Delta U}^{i}$ is the direction of voltage.

So far, we introduce the P&O method at the same environmental condition, i.e., the irradiance and temperature do not change. But environmental condition is always changing over time, it is necessary to operate the PV array at the MPP under varying environmental conditions. Figure 2.22 illustrates how P&O method works when the irradiance changes.

(a) Assume when the irradiance is 900 W/m², the PV array is operating at the MPP, i.e., the point A. When the irradiance changes to 1100 W/m², the PV array is operating at the point B. According to Fig. 2.20, the calculational path ① will be followed, namely in the next step, voltage will be increased to reach point C.

(b) When the voltage increases, we need to compare the current step (the point C) with the previous step (the point B). According to Fig. 2.20, the calculational path ② will be followed, namely in the next step, voltage will be decreased to reach point D.

(c) Then, we need to compare the current step (the point D) with the previous step (the point C). According to Fig. 2.20, the calculational path ③ will be followed, namely in the next step, voltage will be still decreased to reach point E.

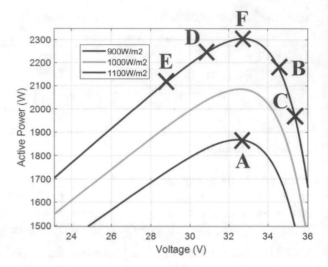

**Fig. 2.22** Illustration of P&O under varying environmental conditions

(d) Then, we will compare the current step (the point E) with the previous step (the point D). According to Fig. 2.20, the calculational path ④ will be followed, namely in the next step, voltage will be increased to reach the MPP, i.e., the point F.

*Note 2.4* From the above four calculation paths, we can see P&O MPPT method is easy to implement, but it may have slow response speed and oscillation around MPP, which can be seen from Example 2.3.

---

**Example 2.3**

As shown in Fig. 2.14, an ideal PV array has 20 PV cell connected in parallel and 40 PV cell in series. The photon current under the reference condition $1000\,\text{W/m}^2$ and $298\,\text{K}$ is $3.35\,\text{A}$. $\eta$ is 1.5, $k$ is $1.38047*10^{-23}\,\text{J/K}$, $q$ is $1.60201*10^{-19}\,\text{C}$. $C_T$ is $2.1775*10^{-5}\,\text{A/K}$, $E_g$ is 1.14, and $I_{sref}$ is $100\,\text{pA}$, i.e., $100*10^{-12}\,\text{A}$.

Question: If the initial operational point is (18.7 V, 1252 W) at $1000\,\text{W/m}^2$ and $298\,\text{K}$, please calculate the first three steps of P&O method when tracking the MPP?

*Case 1*: The increment of voltage at each step is 0.1 V.

*Case 2*: The increment of voltage at each step is 1.0 V.

*Case 3*: The increment of voltage at each step is 8.0 V.

---

Solution:

*Case 1*: Based on (2.18) and (2.19), when the increment of voltage is 0.1 V, according to Fig. 2.20, we can start the iteration process. Figure 2.23 demonstrates

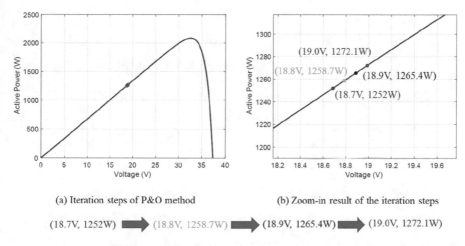

(a) Iteration steps of P&O method  (b) Zoom-in result of the iteration steps

(18.7V, 1252W) ➡ (18.8V, 1258.7W) ➡ (18.9V, 1265.4W) ➡ (19.0V, 1272.1W)

**Fig. 2.23** The first three iterations when the increment of voltage is 0.1 V

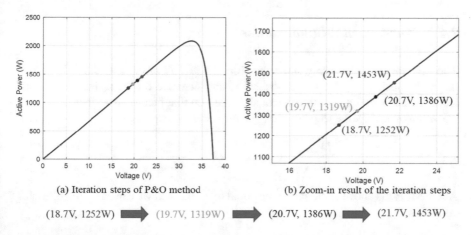

(a) Iteration steps of P&O method  (b) Zoom-in result of the iteration steps

(18.7V, 1252W) ➡ (19.7V, 1319W) ➡ (20.7V, 1386W) ➡ (21.7V, 1453W)

**Fig. 2.24** The first three iterations when the increment of voltage is 1.0 V

the first three steps' iteration, from which we can see when the voltage increment is small, it has very slow response and needs more calculation effort to get to the MPP.

*Case 2*: Based on (2.18) and (2.19), when the increment of voltage is 1.0 V, according to Fig. 2.20, we can start the iteration process. Figure 2.24 demonstrates the first three steps' iteration, from which we can see when the voltage increment is larger than the first case, the iteration process speeds up a bit.

*Case 3*: Based on (2.18) and (2.19), when the increment of voltage is 8.0 V, Fig. 2.25 demonstrates the first three steps' iteration, from which we can see when the voltage increment is very large, we even get an unrealistic result, namely (42.7 V, −86,807 W). It further verifies the voltage increment needs to be carefully selected in practice. ∎

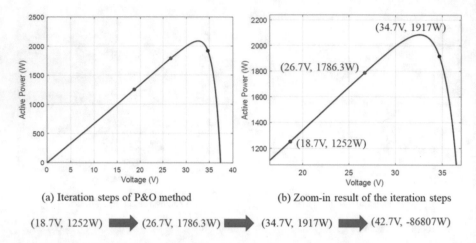

(a) Iteration steps of P&O method                (b) Zoom-in result of the iteration steps

(18.7V, 1252W) ➡ (26.7V, 1786.3W) ➡ (34.7V, 1917W) ➡ (42.7V, -86807W)

**Fig. 2.25** The first three iterations when the increment of voltage is 8.0 V

## 2.6.2   *Incremental Conductance*

The P-U curves of PV array show the slope ($dP/dU$) of the P-U curve is zero at the MPP, positive on the left of the MPP, and negative on the right. Incremental Conductance method is designed exactly based on that fact to track the MPP.

Based on (2.19), the slope of the P-U curve can be expressed as,

$$\frac{dP_{array}}{dU_{array}} = I_{array} + U_{array}\frac{dI_{array}}{dU_{array}}. \tag{2.24}$$

At the MPP, the above slope should be equal to zero. Then we can have the following condition.

$$\frac{dI_{array}}{dU_{array}} = -\frac{I_{array}}{U_{array}}. \tag{2.25}$$

When the PV array is operating on the left of the MPP, we can have (2.26). When the PV array is operating on the right of the MPP, we can have (2.27).

$$\frac{dI_{array}}{dU_{array}} > -\frac{I_{array}}{U_{array}}, \tag{2.26}$$

$$\frac{dI_{array}}{dU_{array}} < -\frac{I_{array}}{U_{array}}. \tag{2.27}$$

From (2.25), (2.26), and (2.27), we can see that by comparing the instantaneous conductance ($I_{array}/U_{array}$) to the incremental conductance ($dI_{array}/dU_{array}$), we

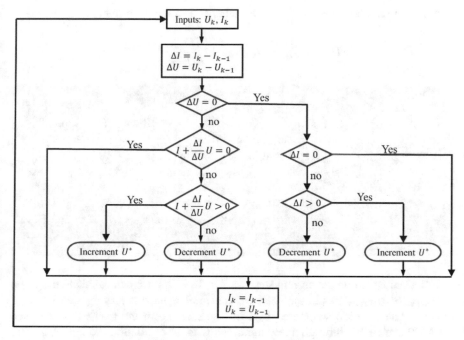

**Fig. 2.26**  Incremental conductance method

can determine how to change the perturbed voltage. Figure 2.26 introduces the Incremental Conductance method.

### 2.6.3   Varying Step MPPT

From the above explanation of the P&O and Incremental Conductance methods, we can see they are both easy to implement, but we need to carefully select the voltage increment to avoid possible oscillation around MPP and speed up the iteration process. To prevent oscillations and meanwhile speeding up the tracking process, a varying step MPPT method is introduced through combing the P&O and incremental conductance methods.

When the PV is operating in the left branch of the P-U curve, we need to increase voltage to get to the MPP. And in the incremental conductance method, the slope ($dP/dU$) is positive, which can be treated as corresponding to voltage increment. When the PV is operating in the right branch of the P-U curve, we need to decrease voltage to get to the MPP. And in the incremental conductance method, the slope ($dP/dU$) is negative, which can also be treated as corresponding to voltage decrements. When the PV is operating in MPP, we do not need to change voltage, and the slope ($dP/dU$) is zero. Therefore, we can use the slope to design a varying step tracking method, as given in (2.28), where $s$ is a scale parameter.

$$U_{k+1,array} = U_{k,array} + \frac{dP_{k,array}}{dU_{k,array}} * s, \tag{2.28}$$

**Example 2.4**
As shown in Fig. 2.14, an ideal PV array has 20 PV cell connected in parallel and 40 PV cell in series. The photon current under the reference condition $1000\,\text{W/m}^2$ and $298\,\text{K}$ is $3.35\,\text{A}$. $\eta$ is 1.5, $k$ is $1.38047 * 10^{-23}\,\text{J/K}$, $q$ is $1.60201 * 10^{-19}\,\text{C}$. $C_T$ is $2.1775 * 10^{-5}\,\text{A/K}$, $E_g$ is 1.14, and $I_{sref}$ is $100\,\text{pA}$, i.e., $100 * 10^{-12}\,\text{A}$.
Question: If the initial operational point is $(18.7\,\text{V}, 1252\,\text{W})$ at $1000\,\text{W/m}^2$ and $298\,\text{K}$, please use the varying step method to track the MPP, where $s$ is 0.03.

Solution: Based on the current step $(18.7\,\text{V}, 1252\,\text{W})$, we can calculate the slope $dP_{k,array}/dU_{k,array}$ in (2.28). Then we can implement the varying step tracking method. The iteration steps are given in Fig. 2.27. From the results, we can see the MPP can be got through a few iterations without oscillation around MPP. The changes of step size during the iterations are summarized in Fig. 2.28, which shows the step size is decreasing as we reach the MPP. ∎

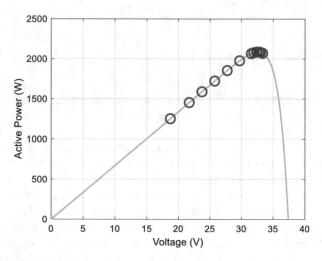

**Fig. 2.27** The iteration steps of varying step MPPT

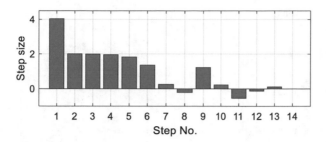

**Fig. 2.28** The changes of step size of varying step MPPT

# Problems

**2.1** Explain how we get current from photovoltaic through the movements of electrons.

**2.2** Explain the impacts of irradiance on the generation of the photovoltaic's photon current $I_{ph}$.

**2.3** Explain the impacts of temperature on the generation of the photovoltaic's photon current $I_{ph}$.

**2.4** An ideal PV array has 30 PV cell connected in parallel and 50 PV cell in series. The photon current under the reference condition $1000\,\text{W/m}^2$ and 298 K is 3.35 A. $\eta$ is 1.5, $k$ is $1.38047 * 10^{-23}$ J/K, $q$ is $1.60201 * 10^{-19}$ C. $C_T$ is $2.1775 * 10^{-5}$ A/K, $E_g$ is 1.14, and $I_{sref}$ is 100 pA, i.e., $100 * 10^{-12}$ A.
   Question 1: What are the open circuit voltage and short circuit current when it is operating under the below circumstances?
*Case 1*: The irradiance is $1000\,\text{W/m}^2$ and the temperature is 298 K.
*Case 2*: The irradiance is $1200\,\text{W/m}^2$ and the temperature is 260 K.
   Question 2: Please roughly draw their output curves, i.e., I-U and P-U, under those two conditions, by using open circuit voltage and short circuit current.

**2.5** An ideal PV array has 80 PV cell connected in parallel and 60 PV cell in series. The photon current under the reference condition $1000\,\text{W/m}^2$ and 298 K is 3.30 A. $\eta$ is 1.5, $k$ is $1.38047 * 10^{-23}$ J/K, $q$ is $1.60201 * 10^{-19}$ C. $C_T$ is $2.1775 * 10^{-5}$ A/K, $E_g$ is 1.14, and $I_{sref}$ is 100 pA, i.e., $100 * 10^{-12}$ A.
   Question 1: What are the open circuit voltage and short circuit current when it is operating under the below circumstances?
*Case 1*: The irradiance is $900\,\text{W/m}^2$ and the temperature is 298 K.
*Case 2*: The irradiance is $1050\,\text{W/m}^2$ and the temperature is 270 K.
   Question 2: Please roughly draw their output curves, i.e., I-U and P-U, under those two conditions, by using open circuit voltage and short circuit current.

**2.6** As shown in Fig. 2.14, an ideal PV array has 20 PV cell connected in parallel and 40 PV cell in series. The photon current under the reference condition

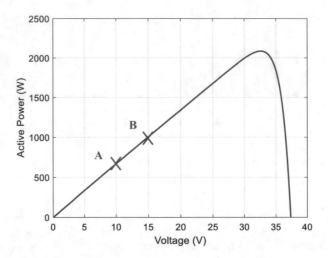

**Fig. 2.29** Example of P&O method for Problem 2.8

1000 W/m$^2$ and 298 K is 3.35 A. $\eta$ is 1.5, $k$ is 1.38047 * 10$^{-23}$ J/K, $q$ is 1.60201 * 10$^{-19}$ C. $C_T$ is 2.1775 * 10$^{-5}$ A/K, $E_g$ is 1.14, and $I_{sref}$ is 100 pA, i.e., 100 * 10$^{-12}$ A.

Question: If the initial operational point is (35.73 V, 1567 W) at 1000 W/m$^2$ and 298 K, please calculate the first three steps of P&O method when tracking the MPP?
*Case 1*: The increment of voltage at each step is 0.2 V.
*Case 2*: The increment of voltage at each step is 2.0 V.
*Case 3*: The increment of voltage at each step is 5.0 V.

**2.7** An ideal PV array has 20 PV cell connected in parallel and 40 PV cell in series. The photon current under the reference condition 1000 W/m$^2$ and 298 K is 3.35 A. $\eta$ is 1.5, $k$ is 1.38047 * 10$^{-23}$ J/K, $q$ is 1.60201 * 10$^{-19}$ C. $C_T$ is 2.1775 * 10$^{-5}$ A/K, $E_g$ is 1.14, and $I_{sref}$ is 100 pA, i.e., 100 * 10$^{-12}$ A. The original PV operational point is (34.88 V, 1872.58 W) at 1000 W/m$^2$ and 298 K.

Question 1: Use P&O method to get the MPP.

Question 2: Give the iteration steps of voltage and power. The voltage increment is given as 0.3 V. The stopping criterion of iteration is 0.5 W.

**2.8** Describe the first five steps on how you use the P&O MPPT strategy to get the maximum power point in Fig. 2.29. Your initial two points are given as point A and point B. Assume your step of P&O is 5 V. Please show your next five steps by using points C, D, E, F, and G.

**2.9** A PV array is originally operating at the point A in Fig. 2.30, to generate more power, P&O MPPT method is used to change the operational point of the PV array. After perturbation, the PV array is observed to operate at the point B.

Question 1: For the first scenario given in Fig. 2.30, which control path in Fig. 2.20 should be adopted next step, ①, ②, ③, or ④? Please explain.

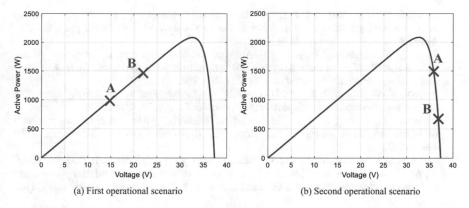

(a) First operational scenario                    (b) Second operational scenario

**Fig. 2.30** Example of P&O method for Problem 2.9

Question 2: For the second scenario given in Fig. 2.30b, which control path should be adopted next step? Please explain.

# References

1. Wenham, S. R., Green, M. A., Watt, M. E., Corkish, R., & Sproul, A. (2013). *Applied photovoltaics*. Routledge.
2. Green, M. A., et al. (2006). *Third generation photovoltaics*. Springer.
3. IEEE standard for interconnection and interoperability of distributed energy resources with associated electric power systems interfaces. *IEEE Std*, 1547–2018.
4. Hanna, M., & Nozik, A. (2006). Solar conversion efficiency of photovoltaic and photoelectrolysis cells with carrier multiplication absorbers. *Journal of Applied Physics, 100*(7), 074510.
5. Topi, M., Brecl, K., & Sites, J. (2007). Effective efficiency of PV modules under field conditions. *Progress in Photovoltaics: Research and Applications, 15*(1), 19–26.
6. Chouder, A., Silvestre, S., Sadaoui, N., & Rahmani, L. (2012). Modeling and simulation of a grid connected PV system based on the evaluation of main PV module parameters. *Simulation Modelling Practice and Theory, 20*(1), 46–58.
7. Ciobotaru, M., Teodorescu, R., & Blaabjerg, F. (2006). Control of single-stage single-phase PV inverter. *Epe Journal, 16*(3), 20–26.
8. Zhao, Z., Xu, M., Chen, Q., Lai, J.-S., & Cho, Y. (2011). Derivation, analysis, and implementation of a boost–buck converter-based high-efficiency PV inverter. *IEEE Transactions on Power Electronics, 27*(3), 1304–1313.
9. Wang, C., Li, Y., Peng, K., Hong, B., Wu, Z., & Sun, C. (2013). Coordinated optimal design of inverter controllers in a micro-grid with multiple distributed generation units. *IEEE Transactions on Power Systems, 28*(3), 2679–2687.
10. Li, Y., Gao, W., & Jiang, J. (2014). Stability analysis of microgrids with multiple der units and variable loads based on MPT. In *2014 IEEE PES General Meeting| Conference & Exposition* (pp. 1–5). IEEE.
11. Agarwal, A., Irtaza, H., & Zameel, A. (2017). Numerical study of lift and drag coefficients on a ground-mounted photo-voltaic solar panel. *Materials Today: Proceedings, 4*(9), 9822–9827.

12. Barkaszi, S. F., & Dunlop, J. P. (2001). Discussion of strategies for mounting photovoltaic arrays on rooftops. In *International Solar Energy Conference* (Vol. 16702, pp. 333–338). American Society of Mechanical Engineers.
13. Pratt, R. N., & Kopp, G. A. (2013). Velocity measurements around low-profile, tilted, solar arrays mounted on large flat-roofs, for wall normal wind directions. *Journal of Wind Engineering and Industrial Aerodynamics*, *123*, 226–238.
14. Trapani, K., & Millar, D. L. (2014). The thin film flexible floating PV (t3f-pv) array: The concept and development of the prototype. *Renewable Energy*, *71*, 43–50.
15. Breazeale, L. C., & Ayyanar, R. (2014). A photovoltaic array transformer-less inverter with film capacitors and silicon carbide transistors. *IEEE Transactions on Power Electronics*, *30*(3), 1297–1305.
16. Liang, T. S., Pravettoni, M., Deline, C., Stein, J. S., Kopecek, R., Singh, J. P., Luo, W., Wang, Y., Aberle, A. G., & Khoo, Y. S. (2019). A review of crystalline silicon bifacial photovoltaic performance characterisation and simulation. *Energy & Environmental Science*, *12*(1), 116–148.
17. Ighneiwa, I., & Yousuf, A. A. (2018). Using intelligent control to improve PV systems efficiency. Preprint. arXiv:1802.03463.
18. Kadri, R., Andrei, H., Gaubert, J.-P., Ivanovici, T., Champenois, G., & Andrei, P. (2012). Modeling of the photovoltaic cell circuit parameters for optimum connection model and real-time emulator with partial shadow conditions. *Energy*, *42*(1), 57–67.
19. Saga, T. (2010). Advances in crystalline silicon solar cell technology for industrial mass production. *NPG Asia Materials*, *2*(3), 96–102.
20. Hentschel, K. (2018). *Photons: The history and mental models of light quanta*. Springer.
21. Moreau, E., Robert, I., Manin, L., Thierry-Mieg, V., Gérard, J., & Abram, I. (2001). Quantum cascade of photons in semiconductor quantum dots. *Physical Review Letters*, *87*(18), 183601.
22. Evans, R. D. (1958). Compton effect. In *Corpuscles and Radiation in Matter II/Korpuskeln und Strahlung in Materie II* (pp. 218–298). Springer.
23. Compton, A. H. (1922). *Secondary radiations produced by X-rays*. National Research Council of the National Academy of Sciences.
24. Meyaard, D. S., Shan, Q., Dai, Q., Cho, J., Schubert, E. F., Kim, M.-H., & Sone, C. (2011). On the temperature dependence of electron leakage from the active region of GaInN/GaN light-emitting diodes. *Applied Physics Letters*, *99*(4), 041112.
25. King, D. L., Kratochvil, J. A., & Boyson, W. E. (1997). Temperature coefficients for pv modules and arrays: measurement methods, difficulties, and results. In *Conference Record of the Twenty Sixth IEEE Photovoltaic Specialists Conference-1997* (pp. 1183–1186). IEEE.
26. Fesharaki, V. J., Dehghani, M., Fesharaki, J. J., & Tavasoli, H. (2011). The effect of temperature on photovoltaic cell efficiency. In *Proceedings of the 1stInternational Conference on Emerging Trends in Energy Conservation–ETEC, Tehran, Iran* (pp. 20–21).
27. Faranda, R., & Leva, S. (2008). Energy comparison of MPPT techniques for PV systems. *WSEAS Transactions on Power Systems*, *3*(6), 446–455.
28. Liu, F., Duan, S., Liu, F., Liu, B., & Kang, Y. (2008). A variable step size INC MPPT method for PV systems. *IEEE Transactions on Industrial Electronics*, *55*(7), 2622–2628.
29. Elobaid, L. M., Abdelsalam, A. K., & Zakzouk, E. E. (2015). Artificial neural network-based photovoltaic maximum power point tracking techniques: a survey. *IET Renewable Power Generation*, *9*(8), 1043–1063.
30. Bendib, B., Belmili, H., & Krim, F. (2015). A survey of the most used MPPT methods: Conventional and advanced algorithms applied for photovoltaic systems. *Renewable and Sustainable Energy Reviews*, *45*, 637–648.

# Chapter 3
# Energy Storage

## 3.1 Battery

Batteries are important to power systems because they give the electric engineer a means for storing small quantities of energy in a way that is immediately available [1]. Applications of batteries includes but not limited to [2–7]:

- Provide the main energy source for electric vehicles
- Function as Uninterruptible Power Supplies (UPS)
- Installed in power grids to adjust active and reactive power
- Used to shave peak power load

There are multiple types of batteries, such as Lead-acid battery [8, 9], Nickel hydrogen battery [10, 11], Nickel cadmium battery [12], Lithium-ion battery [13–15], Sodium-sulfur battery [16, 17], Vanadium redox battery [18, 19], etc.

One important parameter of battery is its capacity [20–22]. Capacity is a measure (typically in Amp-hr) of the charge stored by the battery. It represents the maximum amount of energy that can be extracted from the battery under certain specified conditions. The capacity is determined by the mass of active material contained in the battery, as well as other factors.

State of Charge (SoC) measures the level of charge of an electric battery relative to its capacity [23–25]. The units of SoC are percentage points, i.e., 0% means empty and 100% means full. An alternative form of the same measure is the Depth of Discharge (DoD) [26], which is the inverse of SoC. That is 100% means empty and 0% means full. Therefore, SoC and DoD have the following relation.

$$SoC = 1 - DoD. \tag{3.1}$$

To study the operations of a battery, multiple models can be adopted [27–29], e.g., short-term discharging model, long-term discharging model, over-current battery model, improved battery model, generic model, etc. Short-term discharging model and generic model are introduced as examples.

© Springer Nature Switzerland AG 2022
Y. Li, *Cyber-Physical Microgrids*, https://doi.org/10.1007/978-3-030-80724-5_3

**Fig. 3.1** Short-term
discharging model for battery

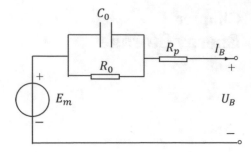

*Short-Term Discharging Model* An example of the short-term discharging model is
given in Fig. 3.1, where $R_p$ is the discharge resistance, and $R_0$ and $C_0$ represent the
over-discharging process. The values of $R_p$, $R_0$, and $C_0$ are changing over time.

According to Fig. 3.1, the battery's output voltage $U_B$ can be calculated in (3.2).

$$U_B(t) = E_m(t) - I_B(t)R_p(t) - U_C(t),  \tag{3.2}$$

where $U_C(t)$ is the voltage over the capacitor and resistor and it can be calculated
using (3.3) according to KCL.

$$C_0(t)\frac{dU_C(t)}{dt} + \frac{U_C(t)}{R_0(t)} = I_B(t).  \tag{3.3}$$

Based on (3.3), we can get the computation of $U_C(t)$, as given in (3.4), where $\tau_0 = R_0(0)C_0(0)$.

$$U_C(t) = I_B(t)R_0(t) - I_B(t)R_0(t)\exp\{-\frac{t}{\tau_0}\}.  \tag{3.4}$$

Therefore, the battery's output voltage $U_B$ can be rewritten as,

$$\begin{aligned} U_B(t) &= E_m(t) - I_B(t)R_p(t) - I_B(t)R_0(t) + I_B(t)R_0(t)\exp\{-\frac{t}{\tau_0}\} \\ &= E_m(t) - I_B(t)R_p(t) - I_B(t)R_0(t)\big(1 - \exp\{-\frac{t}{\tau_0}\}\big). \end{aligned}  \tag{3.5}$$

*Generic Model* When the over-discharging process is ignored, we can get the
generic model. It is widely used to represent a battery. An example is shown in
Fig. 3.2, from which the battery's output voltage $U_B$ can be calculated in (3.6).

$$U_B(t) = E_m(t) - I_B(t)R_p(t).  \tag{3.6}$$

**Fig. 3.2** Generic model for battery

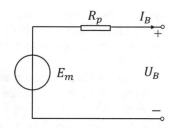

**Fig. 3.3** Typical output of battery voltage

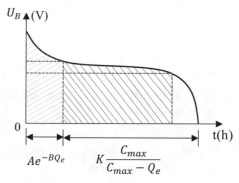

Considering the change of $E_m(t)$ during charging and discharging processes, we usually use (3.7) to calculate it.

$$E_m(t) = E_0 - K \frac{C_{max}}{C_{max} - Q_e} + A \exp\{-B Q_e\}. \qquad (3.7)$$

where,

- $C_{max}$ is the capacity of the battery.
- $Q_e$ is the charge that has been released.
- $K$, $A$, and $B$ are parameters to be determined from experiment.

The typical output voltage of battery $U_B$ is shown in Fig. 3.3. We can see from Fig. 3.3 that there are three operational regions. The first region is mainly determined by $A \exp\{-B Q_e\}$, which shows the beginning of the discharging process. As the battery discharges, the impact of $A \exp\{-B Q_e\}$ will reduce. Since the charge $Q_e$ increases, the impact of $K \frac{C_{max}}{C_{max} - Q_e}$ will increase, which lead to the second region. Battery is mainly operated in this region. In the third region, the output voltage of battery swiftly drops, so we should avoid battery operating in this region.

## 3.2 Flywheel

Flywheel energy storage works by accelerating a rotor (flywheel) to a very high speed and maintaining the energy in the system as rotational energy [30–32]. When energy is extracted from the flywheel, the flywheel's rotational speed will reduce as a consequence of the principle of Conservation of Energy. When energy is added to the flywheel, it is going to correspondingly result in an increase in the speed of the flywheel.

A flywheel is a rotating mechanical device that is used to store rotational energy. The amount of energy stored in a flywheel is proportional to the square of its rotational speed. The energy stored in a flywheel is determined by inertia and rotational speed, which is expressed in (3.8).

$$E = \frac{1}{2} J_F \omega_g^2,$$ (3.8)

where,

- $J_F$ is the inertia, which is also called as the moment of inertia. It resists changes in rotational speed.
- $\omega_g$ is the angular velocity.

Round and ring shape flywheels are frequently used. The inertia for a round shape flywheel is calculated below.

$$J_F = \frac{1}{2} m R^2 = \frac{1}{2} \rho h \pi R^4,$$ (3.9)

where,

- $m$ is the mass.
- $R$ is the radius.
- $h$ is the thickness.
- $rho$ is the density of the flywheel material.

For a ring shape flywheel, the inertia is calculated in (3.10).

$$J_F = \frac{1}{2} m (R_e^2 - R_i^2) - = \frac{1}{2} \rho h \pi (R_e^4 - R_i^4).$$ (3.10)

Based on (3.8), the maximum energy stored in a flywheel is:

$$E_{max} = \frac{1}{2} J_F (\omega_{max}^2 - \omega_{min}^2),$$ (3.11)

where,

- $\omega_{max}$ is the maximum angular velocity.
- $\omega_{min}$ is the minimum angular velocity.

Based on (3.11), we can see there are two ways to increase the energy stored in a flywheel. For the low speed flywheel, we can increase the inertia. For the high speed flywheel, we can increase the maximum angular velocity.

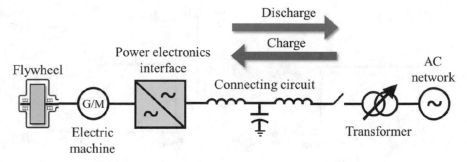

**Fig. 3.4** Charge and discharge of flywheel

When the inertia is fixed, we can charge or discharge power from the flywheel by controlling its angular velocity. The angular velocity and inertia meet the torque calculation as shown in (3.12), where $T$ is the torque of the flywheel. Correspondingly, the power computation is given in (3.13).

$$T = J_F \frac{d\omega_g}{dt}. \tag{3.12}$$

$$P = T\omega_g = J_F\omega_g \frac{d\omega_g}{dt}. \tag{3.13}$$

As an energy storage unit, flywheel can charge and discharge energy, as shown in Fig. 3.4. Flywheel connects to an electric machine. In the charge mode, the electric machine is operating as a motor. In the discharge mode, the electric machine is operating as a generator to produce power. The power electronics interface is used to convert the power energy to nominal frequency, so the flywheel system can integrate to the AC network. The power electronics interface is introduced in Chap. 6.

For the discharge mode, the controller of the power electronics interface is usually used to control the output power, so the flywheel will slow down. For the charge mode, the flywheel will speed up. There are three typical methods [33, 34], namely constant torque method, constant power, and hybrid control.

*Constant Torque Charging Method* In this control, the maximum torque $T_{max\_T}$ is used to speed up the flywheel. Assume in this charging method, the minimum and maximum angular velocities are $\omega_{min\_T}$ and $\omega_{max\_T}$, then the charging power are $T_{max\_T} \cdot \omega_{min\_T}$ and $T_{max\_T} \cdot \omega_{max\_T}$, respectively. The values of flywheel's torque, acceleration, and power are illustrated in Fig. 3.5.

From Fig. 3.5, we can see in the constant torque charging method,

- Since the torque is constant, based on (3.12), the acceleration $\frac{d\omega}{dt}$ is also a constant value, as shown in Fig. 3.5b.
- The angular velocity is calculated through (3.14). Since the acceleration $\left(\frac{d\omega}{dt}\right)_{max\_T}$ is a constant value, the angular velocity is a linear function of

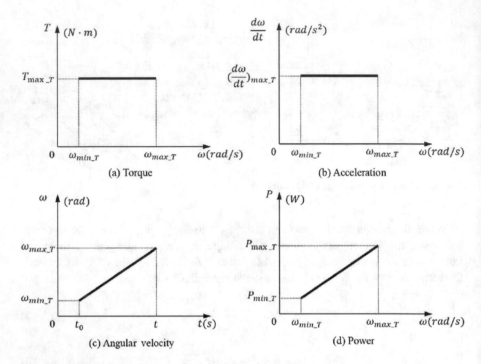

**Fig. 3.5** Constant torque charging for flywheel

the time.

$$\omega_T(t) = \omega_{min\_T} + \int_0^t \left(\frac{d\omega}{dt}\right)_{max\_T} dt = \omega_{min\_T} + \left(\frac{d\omega}{dt}\right)_{max\_T} \cdot t \qquad (3.14)$$

- Since the angular velocity linearly increases, based on (3.13), the power will also linearly increase, as shown in Fig. 3.5d.

The acceleration time is an important factor that needs to be consider in the charging process. The acceleration time can be calculated through (3.14). When the flywheel's angular velocity reaches $\omega_{max}$, we can have the following equation, where $t_T$ is the acceleration time in the constant torque speed up method.

$$\omega_{max\_T} = \omega_{min\_T} + \left(\frac{d\omega}{dt}\right)_{max\_T} \cdot t_T \qquad (3.15)$$

Because $\left(\frac{d\omega}{dt}\right)_{max\_T}$ can be calculated through $\frac{T_{max\_T}}{J_F}$, the acceleration time can be expressed as,

$$t_T = \frac{J_F}{T_{max\_T}}(\omega_{max\_T} - \omega_{min\_T}) \qquad (3.16)$$

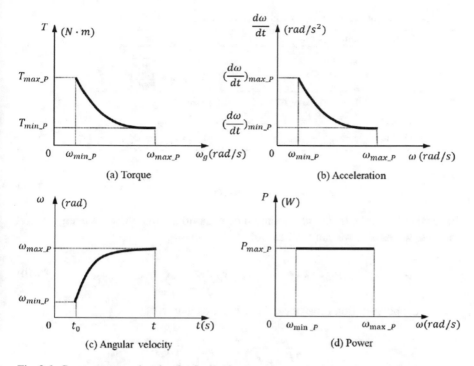

**Fig. 3.6** Constant power charging for flywheel

*Constant Power Charging Method* In this control, the maximum power $P_{max\_P}$ is used to speed up the flywheel, which can be written as $P_{max\_P} = T_P \cdot \omega_P$. Assume in this charging method, the minimum and maximum angular velocities are $\omega_{min\_P}$ and $\omega_{max\_P}$. The values of torque, acceleration, and power are illustrated in Fig. 3.6.

From Fig. 3.6, we can see in the constant power charging method,

- Since the power is constant, based on (3.13), the product of the acceleration $\frac{d\omega}{dt}$ and torque $T$ is a constant value. At the beginning of the charging process, the angular velocity is lower, so the acceleration is high. As the angular velocity increases over time, the acceleration will decrease, as shown in Fig. 3.6b and c.
- Based on (3.12), the changes of torque are same to those of the acceleration, as shown in Fig. 3.6a.
- The angular velocity is calculated through (3.17).

$$\omega_P(t) = \omega_{min\_P} + \int_0^t \frac{d\omega}{dt} dt = \omega_{min\_P} + \int_0^t \frac{P_{max\_P}}{J_F \omega_P} dt \qquad (3.17)$$

Through (3.17), it is difficult to calculate the acceleration time. We will compute the acceleration time from the perspective of the law of the conservation of energy. In the charging mode, the input electric energy can be calculated as,

**Table 3.1** Constant torque method and constant power method

|  | Voltage perturbation | Active power change |
|---|---|---|
| Torque | $T_{max\_T}$ | $T_{min\_P} = \frac{P_{max\_P}}{\omega_{max\_P}}$ <br> $T_{max\_P} = \frac{P_{max\_P}}{\omega_{min\_P}}$ |
| Power | $P_{min\_T} = T_{max\_T} \cdot \omega_{min\_T}$ <br> $P_{max\_T} = T_{max\_T} \cdot \omega_{max\_T}$ | $P_{max\_P}$ |
| Acceleration time | $t_T = \frac{J_F}{T_{max\_T}}(\omega_{max\_T} - \omega_{min\_T})$ | $t_P = \frac{J_F}{2P_{max\_P}}(\omega^2_{max\_P} - \omega^2_{min\_P})$ |

$$E = P_{max\_P} \cdot t_P \tag{3.18}$$

Based on (3.11), the input energy should be equal to the increment of energy stored in the flywheel. So, we can have the following equation.

$$P_{max\_P} \cdot t_P = \frac{1}{2} J_F(\omega^2_{max\_P} - \omega^2_{min\_P}) \tag{3.19}$$

Hence, the acceleration time $t_P$ can be calculated below.

$$t_P = \frac{J_F}{2P_{max\_P}}(\omega^2_{max\_P} - \omega^2_{min\_P}). \tag{3.20}$$

*Note 3.1* The two charging methods are summarized in Table 3.1.

In the two charging methods, the maximum torque $T_{max\_T}$ and $T_{max\_P}$ are equal. So, we can calculate the ratio $K$ of maximum power in these two methods, as given in (3.21).

$$\frac{P_{max\_T}}{P_{max\_P}} = \frac{T_{max\_T} \cdot \omega_{max\_T}}{T_{max\_P} \cdot \omega_{min\_P}} = \frac{\omega_{max\_T}}{\omega_{min\_P}} = K \tag{3.21}$$

Apparently, $K > 1$ when the maximum torque is the same value. So, from (3.21) we can see the maximum power in the constant torque method is larger than that in the constant power method. The ratio of acceleration time can also be calculated, as given in (3.22).

$$\frac{t_T}{t_P} = \frac{\frac{J_F}{T_{max\_T}}(\omega_{max\_T} - \omega_{min\_T})}{\frac{J_F}{2P_{max\_P}}(\omega^2_{max\_P} - \omega^2_{min\_P})} = \frac{2P_{max\_P}}{T_{max\_T}(\omega_{max\_P} + \omega_{min\_P})} \tag{3.22}$$

In (3.22), assume the flywheel is charged from the same state, i.e., the two minimum angular velocities are same $\omega_{min\_T} = \omega_{min\_P}$. According to (3.8), when the flywheel is fully charged, the two maximum angular velocities are also same $\omega_{max\_T} = \omega_{max\_P}$. Since $T_{max\_T} = T_{max\_P}$, (3.22) can be rewritten as,

$$\frac{t_T}{t_P} = \frac{2\omega_{min\_P}}{\omega_{max\_P} + \omega_{min\_P}} = \frac{2}{K+1} < 1 \qquad (3.23)$$

From (3.23), we can see that:

- The constant torque method uses less time to charge the flywheel. Because compared to the constant power method, the acceleration keeps constant to ensure a maximum increment of the flywheel's speed. However, it requires the torque to quickly change to a very small value once the maximum speed reaches.
- The constant power method uses more time to charge the flywheel. The major reason is that the acceleration gradually reduces as the angular velocity increases. So, the flywheel can eventually reach the maximum speed smoothly.

Considering the acceleration time and control systems, a hybrid charging method can incorporate their advantages, as introduced below.

*Hybrid Charging Method* In this control, the constant torque charging method is first used at the beginning of the charging to speed up the process. When the flywheel's speed reaches a threshold $\omega_t$, the constant power charging method is adopted to realize a smooth charging to the maximum speed. The values of torque, acceleration, and power are illustrated in Fig. 3.7.

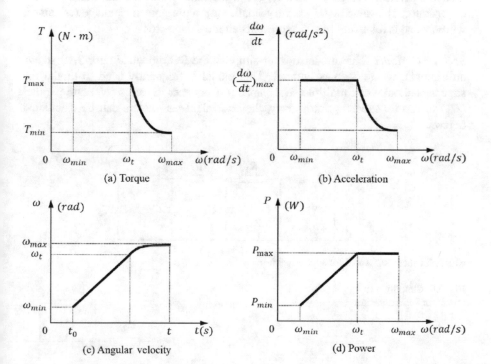

**Fig. 3.7** Hybrid charging for flywheel

## 3.3 Supercapacitor

A supercapacitor, which is also called an ultracapacitor, is a high-capacity capacitor with a capacitance value much higher than other capacitors, but with lower voltage limits [35, 36]. It bridges the gap between electrolytic capacitors and rechargeable batteries.

Supercapacitor includes two carbon electrodes [37]. Physically, the activated carbon electrodes consist of a various size of powder with holes on their respective surfaces. The electrical double layer is formed on the surface where each powder contacts with the electrolyte. Mathematically, supercapacitor can be modeled in multiple ways [38, 39], e.g., linear RC model, simplified RC model, series RC model, etc.

*Linear RC Model* The linear RC model is used to model the performance of the powder with holes on their respective surfaces. The model is given in Fig. 3.8, which shows five time constants. The time constants increase from left to right. This model can represent the detailed performance of the supercapacitor, but it is hard to determine these parameters.

*Simplified RC Model* This model simplify the linear RC model. The simplified model is given in Fig. 3.9, where $R_{ES}$ is a resistance, showing the heating effect of capacitor. The value of $R_{ES}$ is very small. $R_{EP}$ represents the leakage resistance, whose value is relatively big. $C$ represents an ideal capacitor.

*Series RC Model* This model further simplify the RC model. Figure 3.10 shows an example, where $R_{EP}$ is ignored. This model is frequently used to model the supercapacitor. When multiple supercapacitors connect in series ($N_s$) and parallel ($N_p$) to form a supercapacitor array, the equivalent parameters can be calculated below.

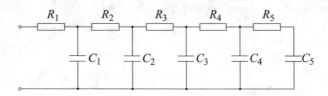

**Fig. 3.8** Linear RC model for supercapacitor

**Fig. 3.9** Simplified RC
model for supercapacitor

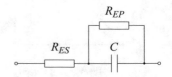

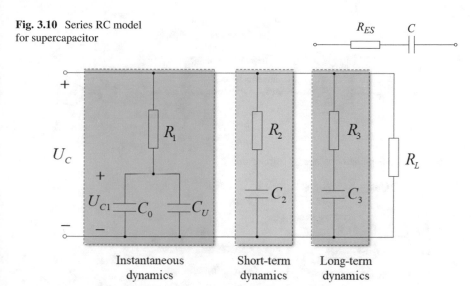

**Fig. 3.10** Series RC model for supercapacitor

**Fig. 3.11** Three-branch model for supercapacitor

$$R_{ES\_array} = \frac{N_s}{N_p} R_{ES} \tag{3.24}$$

$$C_{array} = \frac{N_s}{N_p} C \tag{3.25}$$

The energy stored in a supercapacitor is computed in (3.26).

$$E_C = \frac{1}{2} C U_C^2 \tag{3.26}$$

The simplified RC model and series RC model are easy to analyze the super-capacitor, but these two models only consider the instantaneous dynamics of supercapacitor and do not include the long-term performance. Figure 3.11 shows a three-branch model, which considers the short-term and long-term dynamics. In Fig. 3.11, the left branch models the second level dynamics, the middle branch models the dynamics of a few minutes, and the right branch represents the dynamics in several minutes up to hours.

## 3.4 Applications of Energy Storage Units

Energy storage units have several applications. For instance, they can be used as uninterruptible power supplies (UPS), as shown in Fig. 3.12. When the AC network

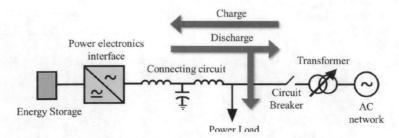

**Fig. 3.12** UPS application of energy storage units

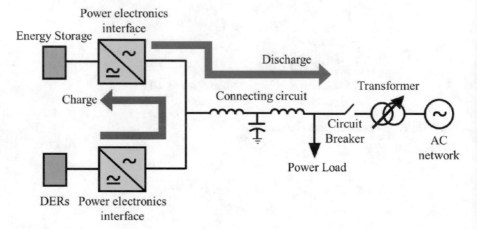

**Fig. 3.13** Smoothing the power outputs of DERs by using energy storage units

is unavailable, the energy storage unit can provide electric power to the load for avoiding blackout. So, it can be installed in power grids for the loads with high priority. Some DERs generate fluctuating power, such as PV and wind. So, energy storage units can also be used to smooth the power outputs of those DERs, as shown in Fig. 3.13. Energy storage units can also be installed in power systems to adjust voltage and frequency by charging or discharging power.

## Problems

**3.1** Describe the function of battery and models for battery.

**3.2** Describe the function of flywheel.

**3.3** Describe the function of supercapacitor and models for supercapacitor.

**3.4** What are the typical charging strategies of flywheel?

**3.5** What are the functions of energy storage units?

# References

1. Oudalov, A., Chartouni, D., & Ohler, C. (2007). Optimizing a battery energy storage system for primary frequency control. *IEEE Transactions on Power Systems, 22*(3), 1259–1266.
2. Mercier, P., Cherkaoui, R., & Oudalov, A. (2009). Optimizing a battery energy storage system for frequency control application in an isolated power system. *IEEE Transactions on Power Systems, 24*(3), 1469–1477.
3. Oudalov, A., Cherkaoui, R., & Beguin, A. (2007). Sizing and optimal operation of battery energy storage system for peak shaving application. In *2007 IEEE Lausanne Power Tech* (pp. 621–625). IEEE.
4. Bhargava, B., & Dishaw, G. (1998). Application of an energy source power system stabilizer on the 10 MW battery energy storage system at chino substation. *IEEE Transactions on Power Systems, 13*(1), 145–151.
5. Oudalov, A., Chartouni, D., Ohler, C., & Linhofer, G. (2006). Value analysis of battery energy storage applications in power systems. In *2006 IEEE PES Power Systems Conference and Exposition* (pp. 2206–2211). IEEE.
6. Lu, C.-F., Liu, C.-C., & Wu, C.-J. (1995). Dynamic modelling of battery energy storage system and application to power system stability. *IEE Proceedings-Generation, Transmission and Distribution, 142*(4), 429–435.
7. Rancilio, G., Lucas, A., Kotsakis, E., Fulli, G., Merlo, M., Delfanti, M., & Masera, M. (2019). Modeling a large-scale battery energy storage system for power grid application analysis. *Energies, 12*(17), 3312.
8. Manwell, J. F., & McGowan, J. G. (1993). Lead acid battery storage model for hybrid energy systems. *Solar Energy, 50*(5), 399–405.
9. Pavlov, D. (2011). *Lead-acid batteries: science and technology*. Elsevier.
10. Zimmerman, A. H. (2009). *Nickel-Hydrogen batteries: Principles and practice*. Aerospace Press.
11. Cai, Z., Liu, G., & Luo, J. (2010). Research state of charge estimation tactics of nickel-hydrogen battery. In *2010 International Symposium on Intelligence Information Processing and Trusted Computing* (pp. 184–187). IEEE.
12. David, J. (1995). Nickel-cadmium battery recycling evolution in Europe. *Journal of Power Sources, 57*(1–2), 71–73.
13. Yoshio, M., Brodd, R. J., & Kozawa, A. (2009). *Lithium-ion batteries* (Vol. 1). Springer.
14. Manthiram, A. (2017). An outlook on lithium ion battery technology. *ACS Central Science, 3*(10), 1063–1069.
15. Gao, L., Liu, S., & Dougal, R. A. (2002). Dynamic lithium-ion battery model for system simulation. *IEEE Transactions on Components and Packaging Technologies, 25*(3), 495–505.
16. Xin, S., Yin, Y.-X., Guo, Y.-G., & Wan, L.-J. (2014). A high-energy room-temperature sodium-sulfur battery. *Advanced Materials, 26*(8), 1261–1265.
17. Bito, A. (2005). Overview of the sodium-sulfur battery for the ieee stationary battery committee. In *IEEE Power Engineering Society General Meeting, 2005* (pp. 1232–1235). IEEE.
18. Skyllas-Kazacos, M., Cao, L., Kazacos, M., Kausar, N., & Mousa, A. (2016). Vanadium electrolyte studies for the vanadium redox battery–a review. *ChemSusChem, 9*(13), 1521–1543.
19. Skyllas-Kazacos, M., & Menictas, C. (1997). The vanadium redox battery for emergency back-up applications. In *Proceedings of Power and Energy Systems in Converging Markets* (pp. 463–471). IEEE.
20. Rong, P., & Pedram, M. (2006). An analytical model for predicting the remaining battery capacity of lithium-ion batteries. *IEEE Transactions on Very Large Scale Integration (VLSI) Systems, 14*(5), 441–451.
21. Casacca, M. A., & Salameh, Z. M. (1992). Determination of lead-acid battery capacity via mathematical modeling techniques. *IEEE Transactions on Energy Conversion, 7*(3), 442–446.

22. Mahdavi-Doost, H., & Yates, R. D. (2013). Energy harvesting receivers: Finite battery capacity. In *2013 IEEE International Symposium on Information Theory* (pp. 1799–1803). IEEE.
23. Chiasson, J., & Vairamohan, B. (2003). Estimating the state of charge of a battery. In *Proceedings of the 2003 American Control Conference, 2003* (Vol. 4, pp. 2863–2868). IEEE.
24. Pop, V., Bergveld, H. J., Notten, P., & Regtien, P. P. (2005). State-of-the-art of battery state-of-charge determination. *Measurement Science and Technology, 16*(12), R93.
25. Charkhgard, M., & Farrokhi, M. (2010). State-of-charge estimation for lithium-ion batteries using neural networks and EKF. *IEEE Transactions on Industrial Electronics, 57*(12), 4178–4187.
26. Duggal, I., & Venkatesh, B. (2014). Short-term scheduling of thermal generators and battery storage with depth of discharge-based cost model. *IEEE Transactions on Power Systems, 30*(4), 2110–2118.
27. Chen, M., & Rincon-Mora, G. A. (2006). Accurate electrical battery model capable of predicting runtime and iv performance. *IEEE Transactions on Energy Conversion, 21*(2), 504–511.
28. Jongerden, M. R., & Haverkort, B. R. (2009). Which battery model to use? *IET Software, 3*(6), 445–457.
29. Tremblay, O., Dessaint, L.-A., & Dekkiche, A.-I. (2007). A generic battery model for the dynamic simulation of hybrid electric vehicles. In *2007 IEEE Vehicle Power and Propulsion Conference* (pp. 284–289). IEEE.
30. Amiryar, M. E., & Pullen, K. R. (2017). A review of flywheel energy storage system technologies and their applications. *Applied Sciences, 7*(3), 286.
31. Genta, G. (2014). *Kinetic energy storage: theory and practice of advanced flywheel systems.* Butterworth-Heinemann.
32. Pena-Alzola, R., Sebastián, R., Quesada, J., & Colmenar, A. (2011). Review of flywheel based energy storage systems. In *2011 International Conference on Power Engineering, Energy and Electrical Drives* (pp. 1–6). IEEE.
33. Buchroithner, A., Wegleiter, H., & Schweighofer, B. (2018). Flywheel energy storage systems compared to competing technologies for grid load mitigation in EV fast-charging applications. In *2018 IEEE 27th International Symposium on Industrial Electronics (ISIE)* (pp. 508–514). IEEE.
34. Li, J., Zhang, H., Wan, Q., Liu, J., & Zhang, H. (2010). A novel charging control for flywheel energy storage system based on BLDC motor. In *2010 Asia-Pacific Power and Energy Engineering Conference* (pp. 1–3). IEEE.
35. Wang, L., Vo, Q.-S., & Prokhorov, A. V. (2017). Stability improvement of a multimachine power system connected with a large-scale hybrid wind-photovoltaic farm using a supercapacitor. *IEEE Transactions on Industry Applications, 54*(1), 50–60.
36. Lee, J.-H., Lee, S.-H., & Sul, S.-K. (2009). Variable-speed engine generator with supercapacitor: Isolated power generation system and fuel efficiency. *IEEE Transactions on Industry Applications, 45*(6), 2130–2135.
37. Iro, Z. S., Subramani, C., & Dash, S. (2016). A brief review on electrode materials for supercapacitor. *International Journal of Electrochemical Science, 11*(12), 10628–10643.
38. Devillers, N., Jemei, S., Péra, M.-C., Bienaimé, D., & Gustin, F. (2014). Review of characterization methods for supercapacitor modelling. *Journal of Power Sources, 246*, 596–608.
39. Zhang, L., Hu, X., Wang, Z., Sun, F., & Dorrell, D. G. (2018). A review of supercapacitor modeling, estimation, and applications: A control/management perspective. *Renewable and Sustainable Energy Reviews, 81*, 1868–1878.

# Chapter 4
# Micro-Turbine

## 4.1 Introduction of Micro-Turbine

Microturbine, system are energy generators whose capacity ranges from 15 kW to 300 kW [1]. Their basic principle comes from the open cycle gas turbines [2, 3]. Microturbines have several features [4–6]: variable speed, high speed operation, compact size, simple operability, easy installation, low maintenance, air bearings, etc. So, they have multiple applications [7], e.g., aircraft, hybrid vehicles, power grids, etc.

Microturbine's physical structure includes a compressor, combustor, turbine and electric generator on a single shaft or two [8]. They can also have a recuperator capturing the waste heat to improve the compressor efficiency. Without recuperator, Microturbines have around 15% efficiencies. With recuperator [9], the efficiency can reach 20–30%. They can reach 85% combined thermal-electrical efficiency in cogeneration. Microturbines rotate at over 30,000 Revolutions Per Minute (RPM). There are two types of microturbines [5]: single-shaft models and double-shaft models.

*Single-Shaft Microturbine* A single expansion turbine turns both the compressor and the generator [10], as shown in Fig. 4.1. A common single-shaft microturbine rotates usually at 45,000–120,000 RPM. Since the electrical power generated by the generator is at a very high frequency (kHz), to integrate microturbine to the AC network, AC-DC-AC power-electronic interfaces are recommended.

The functions of each block in Fig. 4.1 are introduced below.

- Compressor: The inlet air is compressed in a radial (or centrifugal) compressor [11].
- Recuperator: The recuperator uses the heat energy available in the turbine's hot exhaust gas to preheat the compressed air before the compressed air goes into the combustion chamber to increase the overall efficiency [12].

© Springer Nature Switzerland AG 2022
Y. Li, *Cyber-Physical Microgrids*, https://doi.org/10.1007/978-3-030-80724-5_4

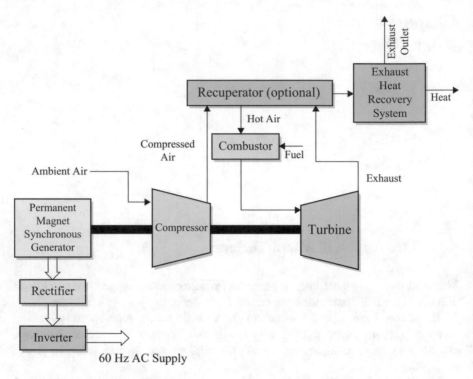

**Fig. 4.1** Single-shaft microturbine

- Combustor: Fuel is mixed with the compressed air in the combustor and burned [13].
- Turbine: The hot combustion gas is then expanded in the turbine section producing rotating mechanical power.
- Permanent Magnet Synchronous Generator: Drive the compressor and the electric generator and generator converts mechanical power to electrical power [14–16].

*Double-Shaft Microturbine* This type of microturbines use a turbine to drive the compressor on one shaft and a power turbine on a separate shaft connected to a conventional generator via a gear box which generates AC power at 60 Hz or 50 Hz. A model is given in Fig. 4.2.

Overall, microturbines are lower power machines with different applications than larger gas turbines, and they have the following typically characteristics [4]:

- Variable rotation: The turbine variable speed is between 30,000 RPM and 120,000 RPM depending on the manufacturer.
- High frequency electric alternator: In the single-shaft microturbine, the generator operates with a converter for AC/DC. In addition, the alternator itself is the engine starter.

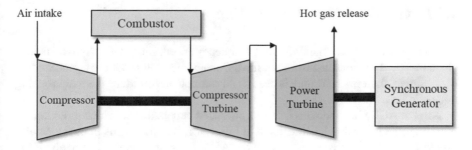

**Fig. 4.2**  Double-shaft microturbine

- Reliability: Microturbines can reach up tp 25,000 h of operation (approximately 3 years) including shutdown and maintenance.
- Simplicity: In the single-shaft microturbine, the generator is placed in the same turbine shaft being relatively easy to be manufactured and maintained. Moreover, it presents a great potential for inexpensive and large-scale manufacturing.
- Compact: Microturbines are easy installation and maintenance.

## 4.2  Synchronous Generator

Synchronous generators are generally used in the microturbine system [17]. A permanent magnet synchronous generator is used in the single-shaft microturbine, which is a special type of synchronous generator. A common synchronous generator is used in the double-shaft microturbine. So, synchronous generator is introduced here. Synchronous generator is commonly used to convert the mechanical power output of gas turbines into electrical power for the grid. The major difference is that the excitation field of the permanent magnet synchronous generator is provided by a permanent magnet instead of a coil.

Stator and rotor are two important components of synchronous generators [18]. For the conventional synchronous generator, the stator is a ring-shaped laminated iron-core with slots. Rotor is DC excitation winding that is wound on an iron core to generate the main magnetic flux. Based on the shape of rotors, there are two types of synchronous generators [18], i.e., round rotor generator and salient pole generator. Salient pole generators are used in hydro power plants. In most cases, hydraulic turbines drive these generators. The round rotor generator is commonly used in fossil and nuclear power plants. Hence, for the microturbine system, round rotor generator is used.

The term synchronous refers here to the fact that the rotor and magnetic field rotate with the same speed. The main rotating magnetic flux, armature reaction, equivalent circuit of synchronous generator are mainly introduced below.

## 4.2.1    Main Rotating Magnetic Flux

Microturbine drives the shaft with high frequency speed. Since DC magnetic field is generated by the rotor, the turning of the DC magnetic field produces a rotating flux $\Phi_{rot}$. Single-shaft microturbine is used as an example to introduce the main rotating magnetic flux.

ABC or ACB phase sequence can be produced depending on which direction the micro-turbine is turning, as illustrated in Fig. 4.3. If the rotor turns in the clockwise direction, then a positive phase sequence (A, B, C) is produced. Conversely, if the rotor spins in the counter-clockwise direction, a negative phase sequence (A, C, B) is generated. In this book, the phase sequence (A, B, C) is used.

As the magnetic flux rotates, the flux linkage with the phase windings varies, as shown in Fig. 4.4. So, we define the angle between the rotating flux vector and the vertical direction as $\omega_r t$, as shown in Fig. 4.5, where $\omega_r$ is the angular velocity of the turbine in the single-shaft microturbine system. Here, the vertical direction is corresponding to the windings of Phase A.

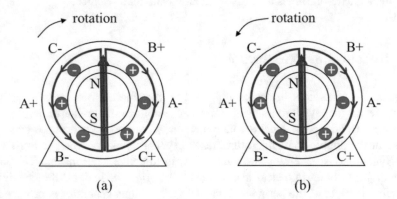

Fig. 4.3   ABC or ACB phase sequence. (a) Clockwise rotation. (b) Counter-clockwise rotation

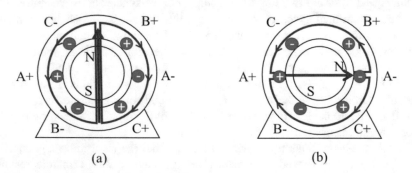

Fig. 4.4   Magnetic flux linkage. (a) Maximum flux linkage with phase A. (b) No flux linkage with phase A

**Fig. 4.5** Changes of angle
when the flux vector rotates

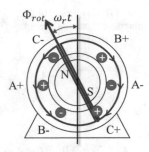

Based on Fig. 4.5, the magnetic flux linkage with phase A can be calculated
through (4.1).

$$\Phi_{link}(t) = \Phi_{rot} \cos(\omega_r t), \tag{4.1}$$

According to Faraday's law of induction [19], any change in the magnetic
environment of a coil of wire will cause a voltage to be induced in the coil. So,
the induced voltage generated in the coil by the magnetic flux is,

$$e(t) = N \frac{d\Phi}{dt}. \tag{4.2}$$

Therefore, substitute (4.1) into (4.2), we can get the induced voltage generated
in the stator. Phase A is used as an example.

$$e_A(t) = N_S \frac{d\Phi_{link}(t)}{dt} = -N_S \frac{d\Phi_{rot} \cos(\omega_r t)}{dt}$$
$$= -N_S \Phi_{rot} \omega_r \sin(\omega_r t) = N_S \Phi_{rot} \omega_r \cos\left(\omega_r t + \frac{\pi}{2}\right). \tag{4.3}$$

From (4.3), we can see the frequency of the stator's induced voltage is $\frac{\omega_r}{2\pi}$, which is
very high as the single-shaft micro-turbine rotates usually at 45,000–120,000 RPM.
The frequency of the induced voltage depends on the turbine's speed. One rotation
generates one sine wave in a two-pole machine. Equation (4.4) shows the RPM
calculation, where $f$ is the frequency, $n_{tur}$ is the turbine's speed, measured by
revolutions per minute, and $p$ is the number of poles.

$$n_{tur} = \frac{120f}{p}. \tag{4.4}$$

From (4.3), we can also calculate the root-mean-square (RMS) value of the induced
voltage $e_A(t)$, that is,

$$E_A = \frac{N_S \Phi_{rot} \omega_r}{\sqrt{2}}. \tag{4.5}$$

**Fig. 4.6** Changes of the magnetic flux and induced voltage over time

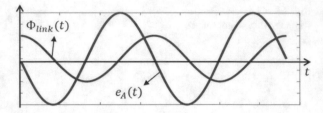

Figure 4.6 illustrates the changes of the magnetic flux and induced voltage over time, from which we can see,

- The induced voltage is leading its corresponding flux linkage 90°.
- The induced voltage in phase A winding will be maximum when the flux linkage is zero, which occurs as the rotor is in parallel with the phase winding.
- The induced voltage is zero when the rotor is perpendicular to the particular phase winding and the flux linkage is maximum.

Note that the phase difference between the voltages generated in the phase windings are 120° because the phase windings are physically placed 120° apart.

## 4.2.2  Armature Reaction

The loading of synchronous generator will produce current in the phase windings. The balanced three-phase load currents produce a rotating (armature) flux with amplitude proportional to the load current. To prove it, we need to first introduce some definitions in the magnetic circuit.

*Flux Density B*  Magnetic flux is the number of magnetic field lines passing through a given closed surface. It provides the measurement of the total magnetic field that passes through a given surface area. The concentration of the magnetic field lines in a magnetic circuit is defined as flux density $B$. It is calculated by,

$$B = \frac{\Phi}{A}, \tag{4.6}$$

where,

- $\Phi$ is the magnetic flux.
- $A$ is the cross-sectional area.

*Magnetomotive Force F*  Magnetomotive force is the driving force that causes a magnetic field. It is defined as,

$$F = N \cdot I, \tag{4.7}$$

where,

- $N$ is the number of turns in the coil.
- $I$ is the magnetizing current supplied to the coil winding generates a magnetic field.

*Magnetic Field Intensity H* Magnetic field intensity is also called field strength. It is the ratio of the magnetomotive force needed to create a certain flux density within a particular material per unit length of that material. The field intensity is defined as,

$$H = \frac{F}{l}, \tag{4.8}$$

where,

- $F$ is the magnetomotive force.
- $l$ is the magnetic path length. The magnetic path length is the average distance traveled by the magnetic flux.

Based on (4.7) and (4.10), we can have the following equation,

$$H \cdot l = N \cdot I. \tag{4.9}$$

*Magnetic Permeability $\mu$* Magnetic permeability is a measure of the ability of a material to carry magnetic flux. It relates flux density and field intensity and is defined as,

$$\mu = \frac{B}{H}. \tag{4.10}$$

Based on the above definitions, we can rewrite the induced voltage,

$$e(t) = N\frac{d\Phi}{dt} = N\frac{dBA}{dt} = AN\frac{d\mu H}{dt}$$

$$= \mu AN\frac{d}{dt}\left(\frac{NI}{l}\right) = \frac{\mu AN^2}{l} \cdot \frac{dI}{dt} = L\frac{dI}{dt}, \tag{4.11}$$

where $L = \frac{\mu AN^2}{l}$ is the inductance created by the coiled winding.

Now, let's prove that the balanced three-phase currents will produce a rotating flux and the amplitude of the rotating flux is proportional to the load current. The rotating flux is called armature flux. From the derivation in (4.11), we can see,

$$\Phi_{arm} = \mu AN\frac{I}{l}. \tag{4.12}$$

Since the load current is AC current, we can express it as $i_{load}(t) = \sqrt{2}I_{load} \cos(\omega_r t + \delta)$. Substitute the load current to (4.12), we can rewrite the magnetic flux,

$$\Phi_{arm} = \sqrt{2}\frac{\mu AN}{l} I_{load} \cos(\omega_r t + \delta). \tag{4.13}$$

From (4.13), we can see the magnetic flux is rotating because of the output high frequency AC load current. Meanwhile, the amplitude of the rotating flux is proportional to the load current. This main rotating flux rotates because rotor is driven by turbine to rotate. This armature flux rotates because AC load current.

Since the magnetic flux is rotating, $\Phi_{arm}$ can be further rewritten as (4.14), where $\Phi_s = \sqrt{2}\frac{\mu AN}{l} I_{load}$.

$$\Phi_{arm} = \Phi_s \cos(\omega_r t + \delta). \tag{4.14}$$

Then, based on the linkage flux (4.1) and armature flux (4.14), the induced voltage generated by the main rotating magnetic flux and armature flux can be summarized in (4.15) and (4.16), respectively.

$$e_A(t) = N_S \Phi_{rot} \omega_r \cos\left(\omega_r t + \frac{\pi}{2}\right). \tag{4.15}$$

$$e_{arm\_A}(t) = N_S \Phi_s \omega_r \cos\left(\omega_r t + \frac{\pi}{2} + \delta\right). \tag{4.16}$$

We can see that (4.15) and (4.16) have the same format, but they are generated in a very different way. The induced voltage $e_A(t)$ is produced due to the rotation of the main magnetic field $\Phi_{rot}$ generated by the rotor. The induced voltage $e_{arm\_A}(t)$ is produced due to the rotation of the armature flux. Their corresponding RMS values are given below.

$$E_A(t) = \frac{N_S \Phi_{rot} \omega_r}{\sqrt{2}}. \tag{4.17}$$

$$E_{arm\_A}(t)\frac{= N_S \Phi_s \omega_r}{\sqrt{2}}. \tag{4.18}$$

### 4.2.3 Equivalent Circuit of Synchronous Generator

A single phase equivalent circuit of synchronous generator is depicted in Fig. 4.7, where $X_{arm}$ is called armature reaction inductance, $X_{leakage}$ represents the stator winding has leakage inductance, and $R_s$ represents the stator coil has resistance.

**Fig. 4.7** Single phase equivalent circuit of synchronous generator

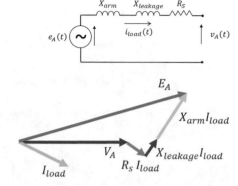

**Fig. 4.8** Example of voltage phasors of a synchronous generator

As introduced in (4.11), the voltage induced by armature flux can be expressed in (4.19), where $i_{load}(t) = \sqrt{2}I_{load}\cos(\omega_r t + \delta)$ is considered.

$$
\begin{aligned}
e_{arm\_A}(t) &= L\frac{di_{load}(t)}{dt} = L\frac{d}{dt}\left(\sqrt{2}I_{load}\cos(\omega_r t + \delta)\right) \\
&= -\sqrt{2}L\omega_r I_{load}\sin(\omega_r t + \delta) \\
&= -\sqrt{2}X_{arm}I_{load}\sin(\omega_r t + \delta) \\
&= \sqrt{2}X_{arm}I_{load}\cos\left(\omega_r t + \frac{\pi}{2} + \delta\right)
\end{aligned}
\tag{4.19}
$$

Based on (4.16) and (4.19), we should have them equal. So, we can get the expression of $X_{arm}$.

$$
X_{arm} = \frac{N_S\Phi_s\omega_r}{\sqrt{2}I_{load}}.
\tag{4.20}
$$

An example of voltage phasors are given in Fig. 4.8 to show their relations.

Therefore, for three-phase balanced synchronous generators, we can eventually express their induced voltage and armature flux induced voltage in the following equations.

$$
\begin{cases}
e_A(t) = N_S\Phi_{rot}\omega_r\cos\left(\omega_r t + \frac{\pi}{2}\right) \\[2mm]
e_B(t) = N_S\Phi_{rot}\omega_r\cos\left(\omega_r t - \frac{\pi}{6}\right) \\[2mm]
e_C(t) = N_S\Phi_{rot}\omega_r\cos\left(\omega_r t - \frac{5\pi}{6}\right),
\end{cases}
\tag{4.21}
$$

$$
\begin{cases}
e_{arm\_A}(t) = N_S \Phi_s \omega_r \cos\left(\omega_r t + \dfrac{\pi}{2} + \delta\right) \\[2ex]
e_{arm\_B}(t) = N_S \Phi_s \omega_r \cos\left(\omega_r t - \dfrac{\pi}{6} + \delta\right) \\[2ex]
e_{arm\_C}(t) = N_S \Phi_s \omega_r \cos\left(\omega_r t - \dfrac{5\pi}{6} + \delta\right).
\end{cases}
\tag{4.22}
$$

**Example 4.1**

A single-shaft turbine drives a four-pole, 250 kVA Y-connected PMSG to generate 1600 Hz voltage. Only 60 kW active power is generated for each phase. The phase-to-phase voltage between phase A and phase B of the generator is $24\angle 30°$ kV. $X_{syn} = X_{arm} + X_{leakage} = 3.0\,\Omega$, $R_S$ is ignored. The number of turns in each phase winding of the stator is 1000.

Question 1: What is the turbine's rotation speed? Please express in RPM.

Question 2: What is the induced stator voltage $E_A$ generated by the main rotating magnetic flux ?

Question 3: What are the main rotating flux $\Phi_{rot}$ and armature flux $\Phi_s$?

Solution: According to (4.4), the RPM can be calculated below.

$$
n_{tur} = \frac{120 f}{p} = \frac{120 \times 1600}{4} = 48{,}000 (RPM).
\tag{4.23}
$$

Because the phase-to-phase voltage between phase A and phase B of the generator is $24\angle 30°$ kV, we can calculate the phase A's voltage is $24\angle 0°/\sqrt{3}$ kV. For each phase, only active power is generated and reactive power is zero, so we can calculate the current for phase A.

$$
I_{load} = \frac{60}{24\angle 0°/\sqrt{3}} = 2.5\sqrt{3}\angle 0° (A).
\tag{4.24}
$$

Then Fig. 4.8 can be leveraged to compute the induced stator voltage $E_A$.

$$
E_A = V_A + j X_{syn} I_{load} = \frac{24\angle 0°}{\sqrt{3}} + 3 * 0.0025\sqrt{3}\angle 90°
$$
$$
= 13.86\angle 0.0537° (kV).
\tag{4.25}
$$

From (4.17) and (4.18), we can get the main rotating flux $\Phi_{rot}$ and armature flux $\Phi_s$. ∎

## 4.3 Power Outputs of Micro-Turbine

It is important to know the power output of microturbine, so it can be used as a dispatchable DERs. Assume the resistance $R_S$ in Fig. 4.7 is ignored, then we can have the following phasor representations, where an inductive load current is used as an example.

In Fig. 4.9, the synchronous generator's output voltage phasor is $E_A \angle \delta_E$ and the integration point's voltage phasor is $V_A \angle \delta_V$. The output current is $I_{load}$. The inductance between the two voltages is $X_{syn}$, which is equal to $X_{arm} + X_{leakage}$. According to KVL, we can calculate the load current, as given in (4.26).

$$I_{load} = \frac{E_A \angle \delta_E - V_A \angle \delta_V}{j X_{syn}}. \tag{4.26}$$

Since $E_A \angle \delta_E$ can be expressed as $E_A \delta_E = E_A(\cos \delta_E + j \sin \delta_E)$ and $V_A \angle \delta_V$ can be expressed as $V_A \angle \delta_V = V_A(\cos \delta_V + j \sin \delta_V)$. Then we can rewrite (4.26) in the following equation.

$$\begin{aligned} I_{load} &= \frac{E_A(\cos \delta_E + j \sin \delta_E) - V_A(\cos \delta_V + j \sin \delta_V)}{j X_{syn}} \\ &= \frac{E_A \cos \delta_E - V_A \cos \delta_V + j(E_A \sin \delta_E - V_A \sin \delta_V)}{j X_{syn}} \\ &= \frac{E_A \sin \delta_E - V_A \sin \delta_V}{j X_{syn}} - j \frac{E_A \cos \delta_E - V_A \cos \delta_V}{j X_{syn}}. \end{aligned} \tag{4.27}$$

So, we can calculate the complex power delivered to the AC system by the synchronous generator.

$$\begin{aligned} S &= V_A \cdot I_{load}^* \\ &= V_A(\cos \delta_V + j \sin \delta_V) \cdot \left( \frac{E_A \sin \delta_E - V_A \sin \delta_V}{j X_{syn}} + j \frac{E_A \cos \delta_E - V_A \cos \delta_V}{j X_{syn}} \right). \end{aligned} \tag{4.28}$$

Based on (4.28), the active power can be expressed as,

**Fig. 4.9** Simplified voltage phasors of a synchronous generator

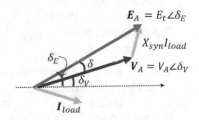

**Fig. 4.10** The relation
between $\delta$ and $P$

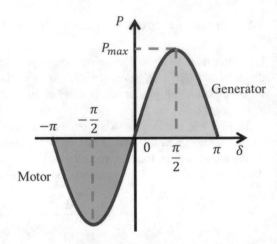

$$P = V_A \cos \delta_V \frac{E_A \sin \delta_E - V_A \sin \delta_V}{j X_{syn}} - V_A \sin \delta_V \frac{E_A \cos \delta_E - V_A \cos \delta_V}{j X_{syn}}$$

$$= \frac{1}{X_{syn}} \left( V_A \cos \delta_V E_A \sin \delta_E - V_A \sin \delta_V E_A \cos \delta_E - V_A \cos \delta_V V_A \sin \delta_V + V_A \cos \delta_V V_A \sin \delta_V \right)$$

$$= \frac{1}{X_{syn}} \left( V_A \cos \delta_V E_A \sin \delta_E - V_A \sin \delta_V E_A \cos \delta_E \right)$$

$$= \frac{1}{X_{syn}} \left( V_A E_A \sin (\delta_E - \delta_V) \right)$$

$$= \frac{1}{X_{syn}} V_A E_A \sin \delta.$$

$$(4.29)$$

where $\sin (\delta_E - \delta_V) = \cos \delta_V \sin \delta_E - \sin \delta_V \cos \delta_E$ is used.

From (4.29), we can see that,

- Since at the steady state, the per unit values of $V_A$ and $E_A$ are around 1.0, the active power output is mainly a function of the angle difference $\delta$. The relation between $\delta$ and $P$ is depicted in Fig. 4.10.
- When $0 \leq \delta \leq \pi$, the active power is positive, i.e., the machine is generating active power; while when $-\pi \leq \delta \leq 0$, the active power is negative, i.e., the machine is using active power and operating in the motor mode.
- Apparently, when $\delta = \pi/2$, the maximum active power output will be produced, assuming that the resistance is negligible. Note this is the single-phase analysis. The maximum power indicated by this equation is called static stability limit of the generator.

If we try to exceed the limit of the generator (such as by admitting more steam to the turbine), the rotor will accelerate. Normally, real generators never even come close to the limit. Full load angle of $15°$–$20°$ are more typical of real generators.

Based on (4.28), the reactive power can be calculated in (4.30).

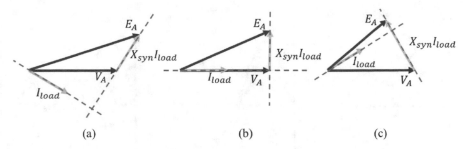

**Fig. 4.11** Determination of power factors. (**a**) Lagging power factor. (**b**) Unity power factor. (**c**) Leading power factor

$$Q = V_A \sin \delta_V \frac{E_A \sin \delta_E - V_A \sin \delta_V}{jX_{syn}} + V_A \cos \delta_V \frac{E_A \cos \delta_E - V_A \cos \delta_V}{jX_{syn}}$$

$$= \frac{1}{X_{syn}} \left( V_A \sin \delta_V E_A \sin \delta_E + V_A \cos \delta_V E_A \cos \delta_E - V_A \sin \delta_V V_A \sin \delta_V - V_A \cos \delta_V V_A \cos \delta_V \right)$$

$$= \frac{1}{X_{syn}} \left( V_A \sin \delta_V E_A \sin \delta_E + V_A \cos \delta_V E_A \cos \delta_E - V_A^2 \right)$$

$$= \frac{1}{X_{syn}} \left( V_A E_A \cos (\delta_E - \delta_V) - V_A^2 \right).$$

$$(4.30)$$

where $\cos (\delta_E - \delta_V) = \sin \delta_V \sin \delta_E + \cos \delta_V \cos \delta_E$ and $\cos (\delta_E - \delta_V) = \sin \delta_V \sin \delta_V + \cos \delta_V \cos \delta_V$ are used.

From (4.30), we can see based on the value of $E_A \cos (\delta_E - \delta_V) - V_A$, we can analyze and determine the power factors, as illustrated in Fig. 4.11 and discussed below.

1. $E_A \cos (\delta_E - \delta_V) - V_A > 0$. In this case, the projection of the synchronous generator's voltage phasor $E_A$ to the $V_A$ is larger than the voltage phasor $V_A$, as illustrated in Fig. 4.11a. So, the reactive power output is positive, namely, the synchronous generator produce reactive power to the AC system.

$$Q = \frac{1}{X_{syn}} \left( V_A E_A \cos (\delta_E - \delta_V) - V_A^2 \right) > 0. \qquad (4.31)$$

In this case, the instantaneous current is lagging voltage, as illustrated in Fig. 4.12a.

2. $E_A \cos (\delta_E - \delta_V) - V_A = 0$. In this case, the projection of the synchronous generator's voltage phasor $E_A$ to the $V_A$ is equal to the voltage phasor $V_A$, as illustrated in Fig. 4.11b. So, the reactive power output is zero, namely, the synchronous generator does not produce reactive power to the AC system.

$$Q = \frac{1}{X_{syn}} \left( V_A E_A \cos (\delta_E - \delta_V) - V_A^2 \right) = 0. \qquad (4.32)$$

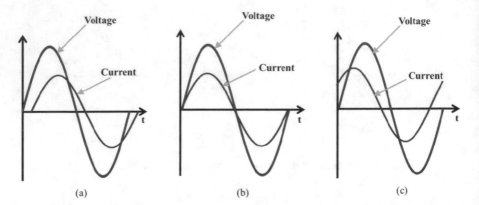

Fig. 4.12 Power factors. (a) Lagging power factor. (b) Unity power factor. (c) Leading power factor

**Fig. 4.13** Test system for Example 2

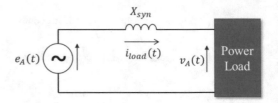

In this case, the instantaneous current is in phase with voltage, as illustrated in Fig. 4.12b.

3. $E_A \cos (\delta_E - \delta_V) - V_A < 0$. In this case, the projection of the synchronous generator's voltage phasor $E_A$ to the $V_A$ is smaller than the voltage phasor $V_A$, as illustrated in Fig. 4.11c. So, the reactive power output is negative, namely, the synchronous generator absorbs reactive power from the AC system.

$$Q = \frac{1}{X_{syn}} \left( V_A E_A \cos (\delta_E - \delta_V) - V_A^2 \right) < 0. \tag{4.33}$$

In this case, the instantaneous current is leading voltage, as illustrated in Fig. 4.12c.

**Example 4.2**
A synchronous generator connects to a power load, as shown in Fig. 4.13. The phase A's voltage of the power load is $V_A = 110\angle 30°$ V, the RMS value of the synchronous generator's induced voltage is 115 V.

(continued)

Question 1: If the power load is a resistive circuit, what is the angle range of the induced voltage $E_A$?

Question 2: If the power load is an equivalent RL circuit, what is the angle range of the induced voltage $E_A$?

Question 3: If the power load is a pure resistive circuit, what is the angle of the induced voltage $E_A$?

Question 4: If the power load is an equivalent RC circuit, what is the angle range of the induced voltage $E_A$?

Question 5: In the given condition, could the generator provide power to a pure inductive circuit? If could, what requirement should be met?

Question 6: In the given condition, could the generator provide power to a pure capacitive circuit? If could, what requirement should be met?

Solution:

*Case 1:* Since the power load is a resistive circuit, it means the active power generated by the synchronous generator must be positive. So, we can have the following inequality,

$$P = \frac{1}{X_{syn}} V_A E_A \sin(\delta_E - \delta_V) = \frac{1}{X_{syn}} V_A E_A \sin\delta > 0. \tag{4.34}$$

Then, we can see when the power load is a resistive circuit, the angle range of the induced voltage $E_A$ is $\delta_E > \delta_V = 30°$.

*Case 2:* Since the power load is an equivalent RL circuit, it means both active power and reactive power generated by the synchronous generator must be positive. So, we can have the following inequality,

$$\begin{cases} P = \dfrac{1}{X_{syn}} V_A E_A \sin(\delta_E - \delta_V) = \dfrac{1}{X_{syn}} V_A E_A \sin\delta > 0 \\ Q = \dfrac{1}{X_{syn}} (V_A E_A \cos(\delta_E - \delta_V) - V_A^2) > 0. \end{cases} \tag{4.35}$$

Then, we can get the following results,

$$\begin{cases} \delta_E > \delta_V = 30° \\ \delta_E < \arccos\left(\dfrac{V_A}{E_A}\right) + \delta_V, \end{cases} \tag{4.36}$$

So, when the power load is an equivalent RL circuit, the angle range of the induced voltage $E_A$ is $30° < \delta_E < 47°$.

*Case 3:* Since the power load is a pure resistive circuit, it means the active power generated by the synchronous generator must be positive and the reactive power must be zero. So, we can have the following equation,

$$
\begin{cases}
P = \dfrac{1}{X_{syn}} V_A E_A \sin(\delta_E - \delta_V) = \dfrac{1}{X_{syn}} V_A E_A \sin\delta > 0 \\
Q = \dfrac{1}{X_{syn}} \left( V_A E_A \cos(\delta_E - \delta_V) - V_A^2 \right) = 0.
\end{cases}
\tag{4.37}
$$

Then, we can get the following results,

$$
\begin{cases}
\delta_E > \delta_V = 30° \\
\delta_E = \arccos\left(\dfrac{V_A}{E_A}\right) + \delta_V,
\end{cases}
\tag{4.38}
$$

So, when the power load is a pure resistive circuit, the angle range of the induced voltage $E_A$ is $\delta_E = 47°$.

*Case 4:* Since the power load is an equivalent RC circuit, it means the active power generated by the synchronous generator must be positive and the reactive power must be negative. So, we can have the following equation,

$$
\begin{cases}
P = \dfrac{1}{X_{syn}} V_A E_A \sin(\delta_E - \delta_V) = \dfrac{1}{X_{syn}} V_A E_A \sin\delta > 0 \\
Q = \dfrac{1}{X_{syn}} \left( V_A E_A \cos(\delta_E - \delta_V) - V_A^2 \right) < 0.
\end{cases}
\tag{4.39}
$$

Then, we can get the following results,

$$
\begin{cases}
\delta_E > \delta_V = 30° \\
\delta_E > \arccos\left(\dfrac{V_A}{E_A}\right) + \delta_V,
\end{cases}
\tag{4.40}
$$

So, when the power load is a pure resistive circuit, the angle range of the induced voltage $E_A$ is $\delta_E > 47°$.

*Case 5:* If we hope the generator provides power to a pure inductive circuit, we need the generator to produce zero active power and the reactive power generated must be positive. So, we can have the following equation,

$$\begin{cases} P = \dfrac{1}{X_{syn}} V_A E_A \sin(\delta_E - \delta_V) = \dfrac{1}{X_{syn}} V_A E_A \sin\delta = 0 \\[2mm] Q = \dfrac{1}{X_{syn}} \left( V_A E_A \cos(\delta_E - \delta_V) - V_A^2 \right) > 0. \end{cases} \tag{4.41}$$

Then, we can get the following results,

$$\begin{cases} \delta_E = \delta_V = 30° \\[2mm] \delta_E < \arccos\left(\dfrac{V_A}{E_A}\right) + \delta_V, \end{cases} \tag{4.42}$$

So, in order to use the generator to energize a pure inductive circuit, we need to adjust the induced voltage' angle to $\delta_E = 30°$.

*Case 6:* If we connect the generator a pure capacitive circuit, we need the generator to produce zero active power and the reactive power generated must be negative because the capacitor will produce reactive power. So, we can have the following equation,

$$\begin{cases} P = \dfrac{1}{X_{syn}} V_A E_A \sin(\delta_E - \delta_V) = \dfrac{1}{X_{syn}} V_A E_A \sin\delta = 0 \\[2mm] Q = \dfrac{1}{X_{syn}} \left( V_A E_A \cos(\delta_E - \delta_V) - V_A^2 \right) < 0. \end{cases} \tag{4.43}$$

Then, we can get the following results,

$$\begin{cases} \delta_E = \delta_V = 30° \\[2mm] \delta_E > \arccos\left(\dfrac{V_A}{E_A}\right) + \delta_V, \end{cases} \tag{4.44}$$

So, in order to use the generator to energize a pure capacitive circuit, from the active power's perspective, we need $\delta_E = 30°$, but from the reactive power's perspective, we need $\delta_E > 47°$. Therefore, in the given condition, the synchronous generator cannot be used to power a capacitor. ∎

From Example 2, we can see when the AC network voltage $V_A$ is maintained as a constant value by the network, mathematically, through changing the angle of $E_A$ we can provide active power and reactive power to different power loads, as discussed in the following cases.

1. The amplitude of the induced voltage $E_A$ is larger than $V_A$ and the amplitude of $E_A$ keeps constant, as illustrated in Fig. 4.14. We can see when the phasor $E_A$ is in the position of $E_A1$, according to (4.29) and (4.30), the active power is zero and reactive power is positive. This case shows the scenario when the power load is a pure inductive circuit. When the phasor $E_A$ is in the position of $E_A2$,

**Fig. 4.14** The amplitude of the induced voltage $E_A$ is larger than $V_A$

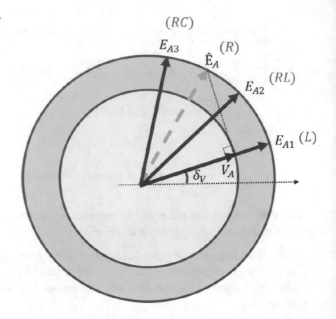

according to (4.29) and (4.30), the active power is positive and reactive power is also positive. This case shows the scenario when the power load is an equivalent RL circuit. When the phasor $E_A$ is in the position of $\hat{E}_A$, according to (4.29) and (4.30), the active power is positive and reactive power is zero. This case shows the scenario when the power load is a pure resistive circuit. When the phasor $E_A$ is in the position of $E_A3$, according to (4.29) and (4.30), the active power is positive and reactive power is negative. This case shows the scenario when the power load is an equivalent RC circuit.

2. The amplitude of the induced voltage $E_A$ is equal to $V_A$ and the amplitude of $E_A$ keeps constant, as illustrated in Fig. 4.15. We can see when the phasor $E_A$ is in the position of $E_{A1}$, according to (4.29) and (4.30), the active and reactive power are zero. This case means it is an open circuit. When the phasor $E_A$ is in the position of $E_{A2}$, according to (4.29) and (4.30), the active power is positive and reactive power is negative. This case shows the scenario when the power load is an equivalent RC circuit. Similar to the case when the phasor $E_A$ is in the position of $E_{A3}$, the power load is an equivalent RC circuit.

3. The amplitude of the induced voltage $E_A$ is smaller than $V_A$ and the amplitude of $E_A$ keeps constant, as illustrated in Fig. 4.16. We can see when the phasor $E_A$ is in the position of $E_{A1}$, according to (4.29) and (4.30), the active power is zero and reactive power is negative. This case shows the scenario when the power load is a pure capacitive circuit. When the phasor $E_A$ is in the position of $E_{A2}$, according to (4.29) and (4.30), the active power is positive and reactive power is negative. This case shows the scenario when the power load is an equivalent RC circuit. Similar to the case when the phasor $E_A$ is in the position of $E_{A3}$, the power load is an equivalent RC circuit.

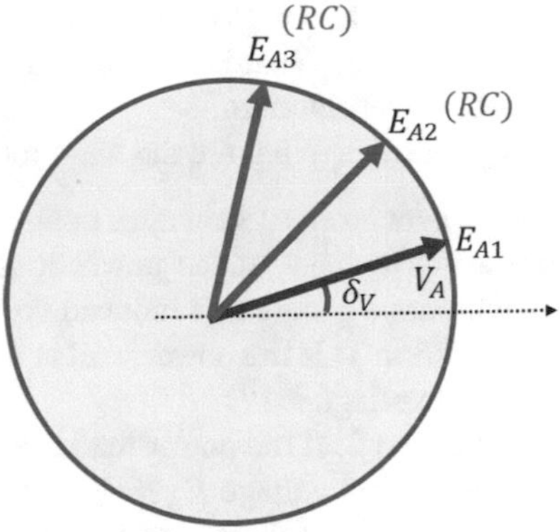

**Fig. 4.15** The amplitude of the induced voltage $E_A$ is equal to $V_A$

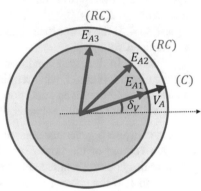

**Fig. 4.16** The amplitude of the induced voltage $E_A$ is smaller than $V_A$

# Problems

**4.1** Describe the function of microturbine.

**4.2** Describe the structure of single-shaft microturbine.

**4.3** Describe the structure of double-shaft microturbine.

**4.4** A single-shaft turbine drives a two-pole, 300 kVA PMSG to generate 800 Hz voltage. Only 100 kW active power is generated for each phase. The phase-to-phase voltage between phase A and phase B of the generator is $20\angle 60°$ kV. $X_{syn} = 1.0\,\Omega$, $R_s$ is ignored. The number of turns in each phase winding of the stator is 2000.

Question 1: What is the turbine's rotation speed? Please express in RPM.

Question 2: What is the induced stator voltage $E_A$ and current $I_{load}$? Please use phasor representation.

Question 3: What are the main rotating flux and armature flux?

**4.5** A single-shaft turbine drives a four-pole, 300 kVA PMSG to generate 1600 Hz voltage. Only 100 kW active power is generated for each phase. The phase-to-phase voltage between phase A and phase B of the generator is $20\angle 60°$ kV. $X_{syn} = 2.0\,\Omega$, $R_s$ is ignored. The number of turns in each phase winding of the stator is 3000.

Question 1: What is the turbine's rotation speed? Please express in RPM.

Question 2: What is the induced stator voltage $E_A$ and current $I_{load}$? Please use phasor representation.

Question 3: What are the main rotating flux and armature flux?

**4.6** A synchronous generator connects to a power load, as shown in Fig. 4.13. The phase A's voltage of the power load is $V_A = 110\angle 30°$V, the RMS value of the synchronous generator's induced voltage is 115 V.

Question 1: If the power load is a resistive circuit, what is the angle range of the induced voltage $E_A$?

Question 2: If the power load is an equivalent RL circuit, what is the angle range of the induced voltage $E_A$?

Question 3: If the power load is a pure resistive circuit, what is the angle of the induced voltage $E_A$?

Question 4: If the power load is an equivalent RC circuit, what is the angle range of the induced voltage $E_A$?

Question 5: In the given condition, could the generator provide power to a pure inductive circuit? If could, what requirement should be met?

Question 6: In the given condition, could the generator provide power to a pure capacitive circuit? If could, what requirement should be met?

**4.7** A synchronous generator connects to a power load, as shown in Fig. 4.13. The phase A's voltage of the power load is $V_A = 110\angle 45°$ V, the RMS value of the synchronous generator's induced voltage is 112 V.

Question 1: If the power load is a resistive circuit, what is the angle range of the induced voltage $E_A$?

Question 2: If the power load is an equivalent RL circuit, what is the angle range of the induced voltage $E_A$?

Question 3: If the power load is a pure resistive circuit, what is the angle of the induced voltage $E_A$?

Question 4: If the power load is an equivalent RC circuit, what is the angle range of the induced voltage $E_A$?

Question 5: In the given condition, could the generator provide power to a pure inductive circuit? If could, what requirement should be met?

Question 6: In the given condition, could the generator provide power to a pure capacitive circuit? If could, what requirement should be met?

# References

1. Peirs, J., Reynaerts, D., & Verplaetsen, F. (2004). A microturbine for electric power generation. *Sensors and Actuators A: Physical, 113*(1), 86–93.
2. Gaonkar, D. & Patel, R. (2006) Modeling and simulation of microturbine based distributed generation system. In *Proceedings of the 2006 IEEE power India conference* (pp. 5–pp). New York: IEEE.

3. Al-Hinai, A. & Feliachi, A. (2002). Dynamic model of a microturbine used as a distributed generator. In *Proceedings of the thirty-fourth Southeastern symposium on system theory (Cat. No. 02EX540)* (pp. 209–213). New York: IEEE.
4. Staunton, R., & Ozpineci, B. (2003) *Microturbine power conversion technology review.* United States: Department of Energy.
5. Soares, C. (2011). *Microturbines: Applications for distributed energy systems.* Amsterdam: Elsevier.
6. Backman, J., & Kaikko, J. (2011). Microturbine systems for small combined heat and power (CHP) applications. In *Small and micro combined heat and power (CHP) systems* (pp. 147–178). Amsterdam: Elsevier.
7. Chahartaghi, M., & Baghaee, A. (2020). Technical and economic analyses of a combined cooling, heating and power system based on a hybrid microturbine (solar-gas) for a residential building. *Energy and Buildings, 217,* 110005.
8. Kiciński, J. & Żywica, G. (2014). *Steam microturbines in distributed cogeneration.* Berlin: Springer.
9. McDonald, C. F. (2003). Recuperator considerations for future higher efficiency microturbines. *Applied Thermal Engineering, 23*(12), 1463–1487.
10. Ofualagba, G. (2012). The modeling and simulation of a microturbine generation system. *International Journal of Scientific & Engineering Research, 2*(7), 1–7.
11. Ashihara, K., Goto, A., Guo, S., & Okamoto, H. (2004). Optimization of microturbine aerodynamics using CFD, inverse design and fem structural analysis: 1st report–compressor design. In *Turbo expo: Power for land, sea, and air*, vol. 41707 (pp. 1513–1519).
12. McDonald, C. F. (2000). Low cost recuperator concept for microturbine applications. In *ASME turbo expo 2000: Power for land, sea, and air.* New York: American Society of Mechanical Engineers Digital Collection.
13. Gieras, M. M. & Stańkowski, T. (2012). Computational study of an aerodynamic flow through a micro-turbine engine combustor. *Journal of Power Technologies, 92*(2), 68–79.
14. Haque, M. E., Negnevitsky, M., & Muttaqi, K. M. (2008). A novel control strategy for a variable speed wind turbine with a permanent magnet synchronous generator. In *Proceeding of the 2008 IEEE industry applications society annual meeting* (pp. 1–8). New York" IEEE.
15. Hasanzadeh, A., Edrington, C. S., Stroupe, N., & Bevis, T. (2013). Real-time emulation of a high-speed microturbine permanent-magnet synchronous generator using multiplatform hardware-in-the-loop realization. *IEEE Transactions on Industrial Electronics, 61*(6), 3109–3118.
16. Wang, T. & Wang, Q. (2012). Optimization design of a permanent magnet synchronous generator for a potential energy recovery system. *IEEE Transactions on Energy Conversion, 27*(4), 856–863.
17. Boldea, I. (2015). *Synchronous generators.* New York: CRC press.
18. Kundur, P. S., Balu, N. J., & Lauby, M. G. (2017). *Power system dynamics and stability.* Boca Raton: CRC Press.
19. Rodrigues, F. G. (2012). On equivalent expressions for the faraday's law of induction. *Revista Brasileira de Ensino de Fisica, 34*(1), 1–6.

# Chapter 5
# Wind Generation

## 5.1 Introduction of Wind Turbines

Wind turbines convert the kinetic energy in the wind to mechanical power [1, 2], where wind is caused by the uneven heating of the earth's surface and rotation of the Earth. Wind turns blades [3, 4], which spin the shaft in a rotor. The rotor spins a generator, which is used to convert the mechanical power into electricity. Therefore, wind turbines use wind to make electricity, which work the opposite of a fan.

A wind turbine turns wind energy into electricity using the aerodynamic force from the rotor blades [5–7]. When wind flows across the blade, the air pressure on one side of the blade decreases. The difference in air pressure across the two sides of the blade creates both lift and drag force. The force of the lift is stronger than the drag, which causes the rotor to spin. The rotor then connects to the generator. Electricity is generated through the translation of the aerodynamic force to the rotation of a generator. Several definitions are introduced below.

*Wind Energy* To analyze the wind generation, we need to first look at the power we get from the wind. Wind power depends on three factors [7], namely amount of air, speed of air, and mass of air. Figure 5.1 is used to analyze the wind power.

The kinetic energy of wind is defined as,

$$E_{wind} = \frac{1}{2} m v_{wind}^2,$$  (5.1)

where,

- $v_{wind}$ is the speed of wind.
- $m$ is the mass of air.

Considering the mass of air can be calculated through (5.2), we can rewrite the wind kinetic energy in (5.3).

© Springer Nature Switzerland AG 2022
Y. Li, *Cyber-Physical Microgrids*, https://doi.org/10.1007/978-3-030-80724-5_5

**Fig. 5.1** Wind power

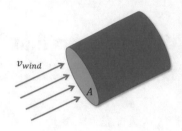

$$m = \rho \pi R_{rotor}^2 v_{wind} t = \rho A v_{wind} t, \tag{5.2}$$

$$E_{wind} = \frac{1}{2} \rho A v_{wind}^3 t, \tag{5.3}$$

where,

– $A$ is the intersection area.
– $\rho$ is the density of the air.
– $t$ is the time.

*Wind Power* Because power is kinetic energy per unit time, we can get the express of wind power through (5.3).

$$P_{wind} = \frac{dE_{wind}}{dt} \frac{1}{2} \rho A v_{wind}^3. \tag{5.4}$$

Considering only a part of the wind power can be captured due to the non-ideal nature of the rotor, the rotor power can be calculated in (5.5), where $C_p$ is the rotor power coefficient [8, 9].

$$P_{rotor} = C_p P_{wind} = \frac{dE_{wind}}{dt} \frac{1}{2} \rho A v_{wind}^3 C_p. \tag{5.5}$$

*Aerodynamic Torque* Based on (5.5), the aerodynamic torque can be developed as follows, where $\omega_{rotor}$ is the rotor's angular velocity.

$$\Gamma_{rotor} = \frac{P_{rotor}}{\omega_{rotor}} = \frac{\rho A v_{wind}^3 C_p}{2\omega_{rotor}}. \tag{5.6}$$

*Tip Speed Ratio (TSR)* $\lambda$ The tip speed ratio is the ratio of the blade-tip linear speed to the wind speed [10, 11], as given in (5.7).

$$\lambda = \frac{\omega_{rotor} R_{rotor}}{v_{wind}}. \tag{5.7}$$

From (5.7), we can see if the rotor's angular velocity $\omega_{rotor}$ is fixed, as the wind speed increases, the tip speed ratio will decrease. If the rotor's angular velocity

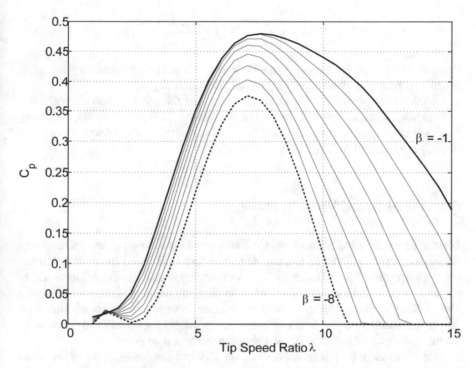

**Fig. 5.2**  Rotor power coefficient $C_p$

$\omega_{rotor}$ is controllable, as the wind speed changes, we can adjust the rotor's angular velocity to keep the tip speed ratio constant. It will be further discussed in the fixed-speed wind turbine and variable-speed wind turbine.

*Rotor Power Coefficient $C_p$*  Rotor power coefficient $C_p$ is a function of tip speed ratio $\lambda$ and the pitch angle $\beta$ [8, 9]. The pitch angle means the angle between the rotation plane of the rotor and the chord line [12]. The function is given in (5.8). Figure 5.2 shows how $\lambda$ and $\beta$ impact $C_p$ [13].

$$C_p = f(\lambda, \beta). \tag{5.8}$$

From Fig. 5.8, we can see that,

- At the same TSR value, as the pitch angle $\beta$ increases, $C_p$ will increase as well.
- For a fixed pitch angle $\beta$, there exists a TSR value, which can have $C_p$ reach its maximum value.

Substitute (5.7) and (5.8) into (5.6), we can update the aerodynamic torque, as follows,

$$\Gamma_{rotor} = \frac{\rho A v_{wind}^3 C_p}{2\omega_{rotor}} = \frac{\rho A v_{wind}^3}{2\omega_{rotor}} f\left(\frac{\omega_{rotor} R_{rotor}}{v_{wind}}, \beta\right). \tag{5.9}$$

From (5.9), we can see the aerodynamic torque is mainly determined by the wind speed $v_{wind}$, rotor's angular velocity $\omega_{rotor}$, and pitch angle $\beta$.

Based on the control of the above variables and the wind system's integration, four typical types of wind turbines are introduced [14], respectively, including fixed-speed wind turbine, variable-slip wind turbine, doubly-fed induction generator (DFIG) turbine, and full converter wind turbine.

## 5.2   Fixed-Speed Wind Turbine

Most utility-scale wind turbines use the fixed-speed type in operations because they are low-cost, robust, reliable, and simple to maintain [15]. This type of wind turbine employs squirrel-cage induction machines to directly connect to the AC network, as shown in Fig. 5.3. For a fixed-speed wind turbine, the pitch-angle control is usually absent. They operate with less than 1% variation in turbine rotor speed. However, due to the uncontrollable generator's speed, the energy captured from the wind is usually sub-optimal and reactive power compensation is required.

From a modeling standpoint, a fixed-speed wind turbine consists of the three major components [14], namely aerodynamic block, mechanical block, and electrical block, as shown in Fig. 5.4. The function of aerodynamic block is to get aerodynamic torque through the wind energy, that is generating kinetic energy. The function of mechanical block is to convert kinetic energy to mechanical energy. And then the electrical block can use mechanical energy to produce electric power. Since

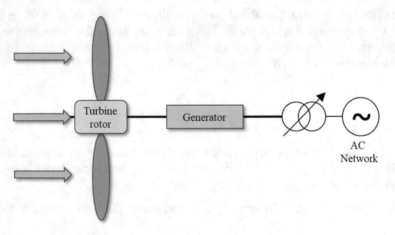

**Fig. 5.3** Fixed-speed wind turbine

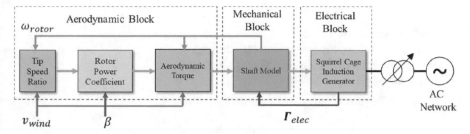

**Fig. 5.4**  Components of fixed-speed wind turbine

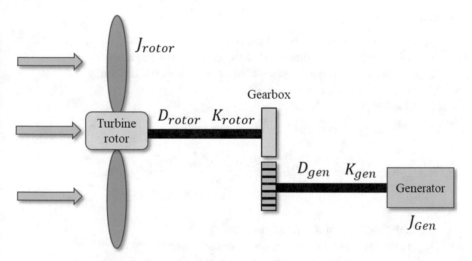

**Fig. 5.5**  Mechanical block of fixed-speed wind turbine

the aerodynamic process has been introduced, we mainly look at the mechanical block and the electrical block.

## 5.2.1   Mechanical Block

The mechanical block consists of the rotor shaft, gearbox, and generator shaft. Two-mass inertia representation shown in Fig. 5.5 is used as an example to model the mechanical block.

For a rotational system consisting of a disk mounted on a shaft fixed at one end, the torque acting on the disk can be calculated by:

$$\Gamma(t) = J\frac{d^2\theta(t)}{dt} + D\frac{d\theta(t)}{dt} + K\theta(t), \tag{5.10}$$

where,

- $J$ is the moment of inertia.
- $D$ is the viscous friction coefficient (damping).
- $K$ is the shaft torsional spring constant (stiffness).
- $\theta(t)$ is the angle.

Then, for the rotor shaft, we can get its torque:

$$\Gamma_{rotor}(t) = J_{rotor}\frac{d^2\theta_{rotor}(t)}{dt} + D_{rotor}\frac{d(\theta_{rotor}(t) - \theta_{gen}(t))}{dt} + K_{rotor}(\theta_{rotor}(t) - \theta_{gen}(t)),$$
(5.11)

where,

- $J_{rotor}$ is the moment of inertia of the rotor.
- $D_{rotor}$ is the viscous friction coefficient (damping) of the rotor.
- $K_{rotor}$ is the shaft torsional spring constant (stiffness) of the rotor.
- $\theta_{rotor}(t)$ is the rotor's angle.
- $\theta_{gen}(t)$ is the generator's angle.

Then, for the generator shaft, we can get its torque:

$$-\Gamma_{gen}(t) = J_{gen}\frac{d^2\theta_{gen}(t)}{dt} + D_{gen}\frac{d(\theta_{gen}(t) - \theta_{rotor}(t))}{dt} + K_{gen}(\theta_{gen}(t) - \theta_{rotor}(t)),$$
(5.12)

where,

- $J_{gen}$ is the moment of inertia of generator.
- $D_{gen}$ is the viscous friction coefficient of the generator.
- $K_{gen}$ is the shaft torsional spring constant of the generator.

### 5.2.2   Induction Generator

Considering the wind speed changes over time, induction generator is usually to convert the mechanical energy to electric power [16, 17]. An induction generator or asynchronous generator is a type of AC electrical generator that uses the principles of induction motors to produce electric power. The output power frequency is determined by the AC system. Induction generator has relatively simple controls.

Induction generator includes stator and rotor [18]. For the stator, the windings are placed in the slots, so we can have two-pole, four-pole, or higher-pole generators. For the rotor, there are mainly two types, namely, squirrel cage rotor and wound rotor. In the squirrel cage rotor, aluminum bus bars are shorted together at the ends by two aluminum rings. This type is most frequently used for wind generation. The wound rotor has conventional three-phase windings made of insulated wire.

**Table 5.1** Synchronous speed

| p | 50 Hz (RPM) | 60 Hz (RPM) |
|---|---|---|
| 2 | 3000 | 3600 |
| 4 | 1500 | 1800 |
| 6 | 1000 | 1200 |
| 8 | 750 | 900 |
| 10 | 600 | 720 |
| 12 | 500 | 600 |
| 30 | 200 | 240 |

An induction generator produces electrical power when its rotor is turned faster than the synchronous speed. Based on (4.4), we can summarize the synchronous speed in Table 5.1, where $p$ is the number of poles.

Since induction generator must rotate faster than the synchronous speed, we can define the relative speed difference between the synchronous speed and the induction generator's speed as slip, which is calculated by,

$$s = \frac{n_{syn} - n_{ind}}{n_{syn}}, \tag{5.13}$$

where,

- $n_{syn}$ is the synchronous speed, as summarized in Table 5.1.
- $n_{ind}$ is the induction generator's speed.

Therefore, we can see for the induction generator, its slip must be smaller than zero, i.e., $s < 0$.

---

**Example 5.1**

In U.S., a fixed-speed wind turbine uses a squirrel-cage induction machines to connect to the AC power grid. Its stator has two poles. The induction generator's speed is 3620 revolutions per minute.

Questions:

Question 1: Is the slip positive or negative?

Question 2: What is the value of the induction generator's slip?

Question 3: If the slip is found within the range of $[-1\%, -0.5\%]$, what is the wind generator's speed range?

---

Solution: As introduced, an induction generator produces electrical power only when its rotor is turned faster than the synchronous speed, so according to (5.13), the slip must be negative.

Since 60 Hz is used in the AC system, we can see the synchronous speed is 3600 RPM, so that we can calculate the slip as below.

$$s = \frac{n_{syn} - n_{ind}}{n_{syn}} = \frac{3600 - 3620}{3600} = -0.556\%. \tag{5.14}$$

When the slip is within the range of $[-1\%, -0.5\%]$, we can have the following inequality,

$$-1\% \leq s = \frac{n_{syn} - n_{ind}}{n_{syn}} \leq -0.5\% \tag{5.15}$$

So that we can calculate the wind generator's speed range is [3618 RPM, 3636 RPM]. ∎

Induction generator has two important operating features about the output of active power and reactive power [19], as introduced below.

- Induction generator cannot generate reactive power. In order to produce electric energy, the induction generator needs to take reactive power from the AC power line. The reason is that reactive power is needed for producing a rotating magnetic field. Because the induction generator absorbs reactive power, if we look at the output voltage and current, the current phasor or trajectory should be always leading the voltage.
- The active power supplied back to the AC system is proportional to the slip above the synchronous speed.

The operation of induction generators is summarized in Fig. 5.6.

*AC Magnetizing Current* The three-phase stator is supplied by the balanced three-phase voltage that drives an AC magnetizing current through each phase winding.

*Magnetic Field* The magnetic field is built up by the AC magnetizing current in the stator. More specifically, the magnetizing current in each phase generates a pulsating AC flux [20]. The flux amplitude varies sinusoidally and the direction of the flux is perpendicular to the phase winding. The three fluxes generated by the phase windings are separated by 120° in space and in time for a two-pole induction generator.

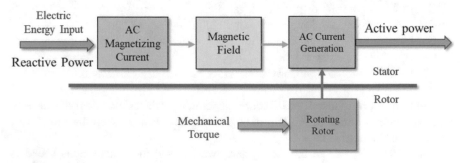

**Fig. 5.6** The operation of induction generators

Based on (4.9) and (4.10), we can get the expression of the pulsating AC flux generated by the AC magnetizing current, as shown in (5.16).

$$
\begin{cases}
\Phi_A(t) = \dfrac{\mu AN}{l} i_A = \sqrt{2}\dfrac{\mu AN}{l} I_A \cos{(\omega t + \delta)} = \hat{\Phi}_A \cos{(\omega t + \delta)} \\[3mm]
\Phi_B(t) = \dfrac{\mu AN}{l} i_B = \sqrt{2}\dfrac{\mu AN}{l} I_B \cos{(\omega t + \delta - \dfrac{2\pi}{3})} = \hat{\Phi}_B \cos{(\omega t + \delta - \dfrac{2\pi}{3})} \\[3mm]
\Phi_C(t) = \dfrac{\mu AN}{l} i_C = \sqrt{2}\dfrac{\mu AN}{l} I_C \cos{(\omega t + \delta + \dfrac{2\pi}{3})} = \hat{\Phi}_C \cos{(\omega t + \delta + \dfrac{2\pi}{3})},
\end{cases}
$$
$$(5.16)$$

where,

- $\Phi_A(t)$ is the instantaneous phase A's magnetic flux.
- $\Phi_B(t)$ is the instantaneous phase B's magnetic flux.
- $\Phi_C(t)$ is the instantaneous phase C's magnetic flux.
- $\hat{\Phi}_A = \sqrt{2}\frac{\mu AN}{l} I_A$ is the amplitude of the phase A's magnetic flux.
- $\hat{\Phi}_B = \sqrt{2}\frac{\mu AN}{l} I_B$ is the amplitude of the phase B's magnetic flux.
- $\hat{\Phi}_C = \sqrt{2}\frac{\mu AN}{l} I_C$ is the amplitude of the phase C's magnetic flux.
- $i_A(t) = \sqrt{2}I_A \cos{(\omega t + \delta)}$ is the phase A's magnetizing current.
- $i_B(t) = \sqrt{2}I_B \cos{(\omega t + \delta - \frac{2\pi}{3})}$ is the phase B's magnetizing current.
- $i_C(t) = \sqrt{2}I_C \cos{(\omega t + \delta + \frac{2\pi}{3})}$ is the phase C's magnetizing current.

From (5.16), we can see each phase's magnetic field is an AC flux, changing at synchronous speed with its amplitude proportional to the magnetizing current's amplitude [21]. They are separated by 120° in time. Since the three fluxes are also separated by 120° in space, we can use Fig. 5.7 to illustrate the three fluxes.

The total flux is the sum of the three fluxes. In Fig. 5.7, considering the space separation, the total magnetic flux can be calculated in (5.17).

**Fig. 5.7** Illustration of three magnetic fluxes

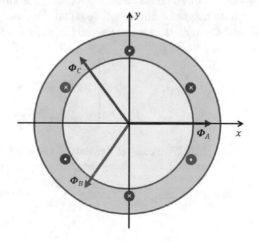

$$\Phi_{total} = \Phi_A \angle 0° + \Phi_B \angle - 120° + \Phi_C \angle 120°$$

$$= \hat{\Phi}_A \cos(\omega t + \delta)\angle 0° + \hat{\Phi}_B \cos(\omega t + \delta - \frac{2\pi}{3})\angle - 120° \qquad (5.17)$$

$$+ \hat{\Phi}_C \cos(\omega t + \delta + \frac{2\pi}{3})\angle 120°$$

Projecting the three magnetic fluxes to $x$-axis and $y$-axis, respectively, we can then get,

$$\begin{cases} \Phi_{total\_x} = \hat{\Phi}_A \cos(\omega t + \delta)\cos(0°) + \hat{\Phi}_B \cos(\omega t + \delta - \frac{2\pi}{3})\cos(-120°) \\ \qquad + \hat{\Phi}_C \cos(\omega t + \delta + \frac{2\pi}{3})\cos(120°) \\ \Phi_{total\_y} = \hat{\Phi}_A \cos(\omega t + \delta)\sin(0°) + \hat{\Phi}_B \cos(\omega t + \delta - \frac{2\pi}{3})\sin(-120°) \\ \qquad + \hat{\Phi}_C \cos(\omega t + \delta + \frac{2\pi}{3})\sin(120°) \end{cases}$$

$$(5.18)$$

Then through calculation, we can have,

$$\begin{cases} \Phi_{total\_x} = \hat{\Phi}_A \cos(\omega t + \delta) + \hat{\Phi}_B \cos(\omega t + \delta - \frac{2\pi}{3})(-\frac{1}{2}) \\ \qquad + \hat{\Phi}_C \cos(\omega t + \delta + \frac{2\pi}{3})(-\frac{1}{2}) \\ \Phi_{total\_y} = \hat{\Phi}_B \cos(\omega t + \delta - \frac{2\pi}{3})(-\frac{\sqrt{3}}{2}) + \hat{\Phi}_C \cos(\omega t + \delta + \frac{2\pi}{3})(\frac{\sqrt{3}}{2}) \end{cases}$$

$$(5.19)$$

For the three-phase balanced AC magnetizing current, the amplitude of the three-phase magnetic fluxes will be equal, i.e., $\hat{\Phi}_A = \hat{\Phi}_B = \hat{\Phi}_C = \hat{\Phi}$. Replace $\hat{\Phi}_A$, $\hat{\Phi}_B$, and $\hat{\Phi}_C$ with $\hat{\Phi}$, we can rewrite (5.19) in the following expression.

$$\begin{cases} \Phi_{total\_x} = \frac{3\hat{\Phi}}{2}\cos(\omega t + \delta) \\ \Phi_{total\_y} = \frac{3\hat{\Phi}}{2}\sin(\omega t + \delta) \end{cases} \qquad (5.20)$$

From (5.20), we can get the sum of the three fluxes,

$$\Phi_{total} = \Phi_{total\_x} + j\Phi_{total\_y} = \frac{3\hat{\Phi}}{2}\cos(\omega t + \delta) + j\frac{3\hat{\Phi}}{2}\sin(\omega t + \delta). \qquad (5.21)$$

**Table 5.2**  Comparison between induction generator and PMSG

|  | Permanent Magnetic Synchronous Generator | Induction Generator |
|---|---|---|
| Main magnetic field | Rotor | Stator |
| Generation of main magnetic field | Permanent magnet | Stator (AC current) |
| Rotating field | Rotor is spinning | AC current |

By using Euler's Identify $e^{j\theta} = \cos\theta + j\sin\theta$, we can get the following total magnetic flux,

$$\Phi_{total} = \frac{3\hat{\Phi}}{2}e^{j(\omega t+\delta)}. \qquad (5.22)$$

From (5.22), we can see that,

- The summation of the three AC fluxes results in a rotating flux.
- When the frequency of the magnetizing current does not change, the rotating total magnetic flux turns with the constant speed $\omega$.
- When the amplitude of the magnetizing current does not change, the rotating total magnetic flux turns with constant amplitude $\frac{3\hat{\Phi}}{2}$.

From the above derivation, we can see the induction generator is very different from the permanent magnetic synchronous generator introduced before. Their major differences are summarized in Table 5.2.

*Rotating Rotor and Generating AC Current*  For the fixed-speed wind turbine, it operates with less than 1% variation in the turbine rotor's speed, i.e., $\omega_{rotor}$ is less than 1% variation. From (5.7), we can see when $\omega_{rotor}$ is fixed, the value of TSR is only determined by the wind speed $v_{wind}$.

In Fig. 5.8, when the pitch-angle control is absent, i.e., $\beta$ does not change, we can have the following findings.

- There exists a TSR value, which can have $C_p$ reach its maximum value.
- In the right branch of $C_p$'s maximum value, as the wind speed $v_{wind}$ increases, $\lambda$ will decrease and $C_p$ will also increase. Based on (5.5), the increase of $v_{wind}$ and $C_p$ will lead to an increase of $P_{rotor}$. In other words, when the wind speed is relatively small, intuitively, the increase of the wind speed will result in more power output.
- In the left branch of $C_p$'s maximum value, as the wind speed $v_{wind}$ continues increasing, $\lambda$ will decrease as well as $C_p$. The increase of $v_{wind}$ and decrease of $C_p$ will lead to a change of $P_{rotor}$. In other words, when the wind speed is over a threshold, the increase of wind speed may not result in more power output due to the decrease of $C_p$.

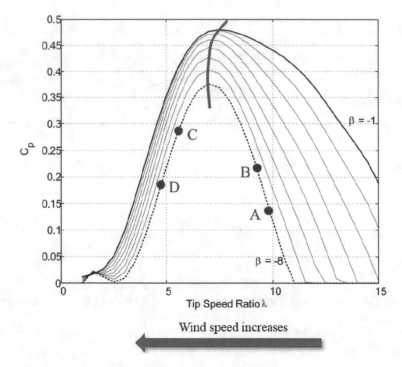

**Fig. 5.8** Tip speed ratio analysis

Figure 5.9 shows the changes of power output as the wind speed and rotor's angle velocity vary.

From Fig. 5.9, we can get the following general findings that are applied to all types of wind turbines.

- For each specific curve, namely with the same wind speed, there exists a rotor's angular velocity that can have the induction generator produce the maximum power.
- For the same rotor's angular velocity, different wind speed will lead to different power output.
- There is an optimal power curve for different wind speeds.

For the fixed-speed wind turbine, the rotor's angular velocity $\omega_{rotor}$ is fixed, then we can see that,

- When the rotor's angular velocity is fixed, the generator can only produce the maximum power at one specific wind speed. For instance, when the wind speed is $v_{wind7}$ and the rotor's angular velocity is $\omega_{rotorA}$, the induction generator will produce the maximum power, as shown by point A.
- The induction generator usually produces sub-optimal power output. For instance, the rotor's angular velocity is fixed at $\omega_{rotorA}$, when the wind speed is

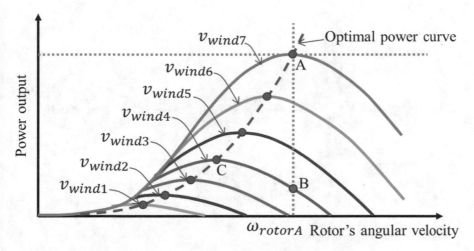

**Fig. 5.9**  Power outputs when the wind speed and rotor's angle velocity vary

$v_{wind4}$, the induction generator only produce power shown by point B; however, at this wind speed, the maximum power is point C, which is much larger than the power at point B.

## 5.3   Variable-Slip Wind Turbine

Fixed-speed wind turbine cannot optimally extract power from the wind. Under this background, variable-slip wind turbine is designed to operate at a wide range of rotor speeds. These turbines usually employ blade-pitching. Speed and power controls allow these turbines to extract more energy from a given wind regime than fixed-speed turbines can. But some power is lost as heat in the rotor resistance because of variable-slip or dynamic rotor resistance turbines control. A variable-slip wind turbine is given in Fig. 5.10.

The major difference between the fixed speed wind turbine and variable-slip wind turbine are summarized in Table 5.3.

Figure 5.11 illustrates the power outputs by the variable-slip wind turbine, from which we can see that since the rotor's angular velocity can adjustable, the induction generator can produce the maximum power output at different wind speed. For instance, assume the initial rotor's angular velocity is $\omega_{rotor7}$ to extract the maximum power at the wind speed $v_{wind7}$, when the wind speed changes to $v_{wind4}$, the rotor's angular velocity will correspondingly change to $\omega_{rotor4}$. So, the generator can produce the maximum power shown by point C, instead of the power at point B produced by the fixed-speed wind turbine.

Therefore, theoretically, the variable-slip wind turbine can adjust its rotor's angular velocity to guarantee the generator always extracts the maximum power.

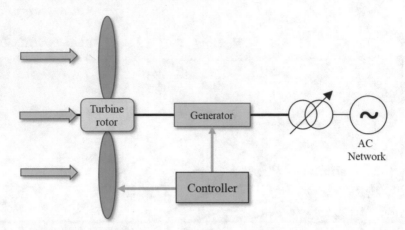

**Fig. 5.10**  Variable-slip wind turbine

**Table 5.3**  Comparison between the fixed speed wind turbine and variable-slip wind turbine

|                      | Fixed-Speed Wind Turbine                          | Variable-Slip Wind Turbine |
|----------------------|---------------------------------------------------|----------------------------|
| Induction Generator  | Squirrel cage                                     | Wound-rotor                |
| Rotor speed          | Fixed                                             | A wide range of speeds     |
| Energy               | Energy capture from the wind is sub-optimal       | Maxmium energy             |

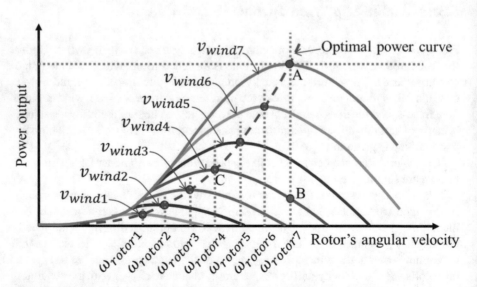

**Fig. 5.11**  Power outputs of the variable-slip wind turbine

## 5.4 Doubly-Fed Induction Generator (DFIG) Wind Turbine

To reduce the power loss in the variable-slip wind turbine, Doubly-Fed induction generator (DFIG) turbines are designed to remedy the power loss problem by employing a back-to-back AC/DC/AC converter in the rotor circuit to recover the slip power. DFIG has emerged as the generator technology of choice for modern wind farms. Flux-vector control of rotor currents allows decoupled active and reactive power output, as well as maximized wind power extraction and lowering of mechanical stresses [22–24]. A typical DFIG wind turbine is given in Fig. 5.12.

In DFIG turbines, A wound-rotor induction machine is used to generator electric power. A back-to-back AC-DC-AC power electronic converter is used to rectify the supply voltage and convert it to three-phase AC at the desired frequency for rotor excitation. Unlike a singly-excited squirrel-cage induction machine, stator and rotor windings of a DFIG are independently excited.

Since the induction generator has been introduced in the fixed-speed wind turbine, here we mainly look at the frequency of voltage and current generated in the rotor side.

According to Faraday's Law of Induction, a voltage is induced in a circuit whenever relative motion exists between a conductor and a magnetic field and that the magnitude of this voltage is proportional to the rate of change of the flux. Since the magnetic field has synchronous speed as derived in (5.22), the rotor's speed changes will cause the rotor's voltage's and current's frequency to change.

One simulation example given in Example 2 is used to demonstrate the frequency changes when rotor's speed changes.

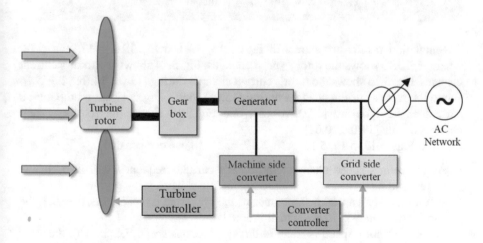

**Fig. 5.12** Doubly-Fed Induction Generator (DFIG)

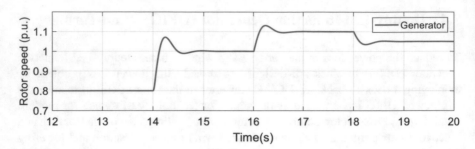

**Fig. 5.13** Rotor's speed changes

**Example 5.2**
A DFIG is connected to the AC network through back-to-back power electronics interface given in Fig. 5.12. Four operation scenarios are under investigation.
*Case 1*: During $[12.0s, 14.0s]$, the rotor's speed is 0.80 times of the synchronous speed.
*Case 2*: During $[14.0s, 16.0s]$, the rotor's speed is equal to the synchronous speed.
*Case 3*: During $[16.0s, 18.0s]$, the rotor's speed is 1.10 times of the synchronous speed.
*Case 4*: During $[18.0s, 20.0s]$, the rotor's speed is 1.05 times of the synchronous speed.
Questions: Please demonstrate the rotor's current changes.

Simulation results are shown in Figs. 5.13, 5.14, 5.15, 5.16, 5.17, and 5.18, where Fig. 5.13 shows the rotor's speed changes, Fig. 5.14 shows the rotor's current changes, Fig. 5.15 shows the rotor's current changes during $[13.0s, 14.0s]$, Fig. 5.16 shows the rotor's current changes during $[15.0s, 16.0s]$, Fig. 5.17 shows the rotor's current changes during $[17.0s, 18.0s]$, and Fig. 5.18 shows the rotor's current changes during $[19.0s, 20.0s]$.

From Figs. 5.13, 5.14, 5.15, 5.16, 5.17, and 5.18, we can see that,

- As the rotor's speed changes, the rotor's current frequency will also change correspondingly.
- When the rotor's speed is 0.80 times of the synchronous speed, i.e., $s = 0.2$, the current frequency is 12 Hz, which is equal to $s \cdot f_n$.
- When the rotor's speed is equal to the synchronous speed, i.e., $s = 0$, the rotor generators DC current, that is the frequency is zero, which is equal to $s \cdot f_n$.
- When the rotor's speed is 1.10 times of the synchronous speed, i.e., $s = -0.1$, the current frequency is 6 Hz, which is equal to $-s \cdot f_n$.

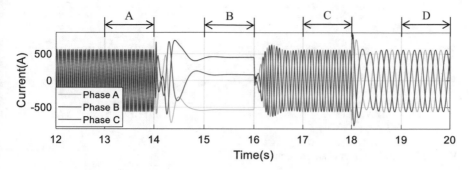

**Fig. 5.14** Rotor's current changes between 12s and 20s

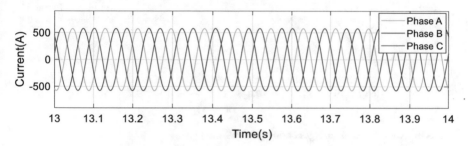

**Fig. 5.15** Rotor's current changes during $[13.0s, 14.0s]$

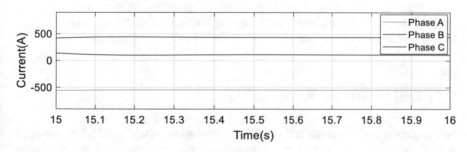

**Fig. 5.16** Rotor's current changes during $[15.0s, 16.0s]$

- When the rotor's speed is 1.05 times of the synchronous speed, i.e., $s = -0.05$, the current frequency is 3 Hz, which is equal to $-s \cdot f_n$. ∎

From the analysis of Example 2, we can see the frequency of current generated in the rotor can be calculated by (5.23).

$$f_{rotor} = |s| f_n. \tag{5.23}$$

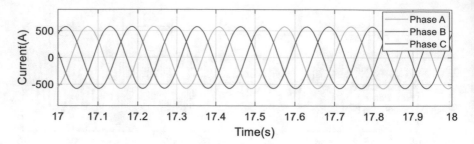

**Fig. 5.17** Rotor's current changes during [17.0s, 18.0s]

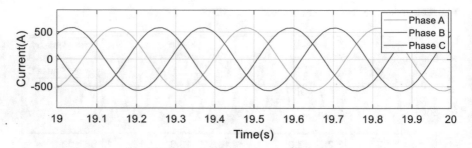

**Fig. 5.18** Rotor's current changes during [19.0s, 20.0s]

Note that (5.23) shows the frequency of current generated in the rotor depends on the absolute value of slip, which means theoretically, there are two speeds, at which the wind generator can produce rotor's current with the same frequency.

## 5.5   Full Converter Wind Turbine

Full converter wind turbine has no direct connection to the grid. A back-to-back AC/DC/AC converter is the only power flow path from the wind turbine to the grid [23, 25]. These turbines may employ synchronous or induction generators and offer independent active and reactive power control. A full converter wind turbine is shown in Fig. 5.19.

## Problems

**5.1** What is the function of wind generation?

**5.2** What are four typical types of wind turbines?

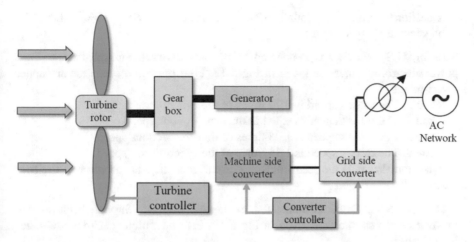

**Fig. 5.19** Full converter wind turbine

**5.3** Explain the major difference between permanent magnetic synchronous generator and the induction generator.

**5.4** Explain the major difference between the fixed speed wind turbine and variable-slip wind turbine.

**5.5** Explain how the power output is controlled in the type of variable-slip wind turbines.

**5.6** What is slip? For the induction generator, is slip a positive value, negative, or zero?

**5.7** In the induction generator, please explain the total magnetic field is AC flux, changing at synchronous speed with the amplitude proportional to the magnetizing current's amplitude.

**5.8** In U.S., a fixed-speed wind turbine uses a squirrel-cage induction machines to connect to the AC power grid. Its stator has four poles. The induction generator's speed is 1810 revolutions per minute.
   Question 1: Is the slip positive or negative?
   Question 2: What is the value of the induction generator's slip?
   Question 3: If the slip is found within the range of [−0.8%, −0.5%], what is the wind generator's speed range?

**5.9** In U.S., a fixed-speed wind turbine uses a squirrel-cage induction machines to connect to the AC power grid. Its stator has eight poles. The induction generator's speed is 910 revolutions per minute.
   Question 1: Is the slip positive or negative?
   Question 2: What is the value of the induction generator's slip?

Question 3: If the slip is found within the range of $[-1.2\%, -0.3\%]$, what is the wind generator's speed range?

**5.10** In, U.S., a DFIG is connected to the AC network through back-to-back power electronics interface given in Fig. 5.12. Four operation scenarios are under investigation.

*Case 1*: The rotor's speed is 0.70 times of the synchronous speed.
*Case 2*: The rotor's speed is equal to the synchronous speed.
*Case 3*: The rotor's speed is 1.05 times of the synchronous speed.
*Case 4*: The rotor's speed is 1.15 times of the synchronous speed.

Question: What is the frequency of current generated in the rotor in the four cases?

**5.11** In, U.S., a DFIG is connected to the AC network through back-to-back power electronics interface given in Fig. 5.12. Five operation scenarios are under investigation.

*Case 1*: The frequency of current generated in the rotor is 0 Hz.
*Case 2*: The frequency of current generated in the rotor is 10 Hz.
*Case 3*: The frequency of current generated in the rotor is 20 Hz.
*Case 4*: The frequency of current generated in the rotor is 25 Hz.
*Case 5*: The frequency of current generated in the rotor is 30 Hz.

Question: What are the rotor's speeds in the above five cases?

# References

1. Spera, D. A. (1994). *Wind turbine technology*.
2. Bossart, S. (2012). DOE perspective on microgrids. In *Advanced microgrid concepts and technologies workshop*.
3. Mishnaevsky, L., Branner, K., Petersen, H. N., Beauson, J., McGugan, M., & Sørensen, B. F. (2017). Materials for wind turbine blades: An overview. *Materials, 10*(11), 1285.
4. Xudong, W., Shen, W. Z., Zhu, W. J., Sørensen, J. N., & Jin, C. (2009). Shape optimization of wind turbine blades. *Wind Energy: An International Journal for Progress and Applications in Wind Power Conversion Technology, 12*(8), 781–803.
5. Lee, K., Huque, Z., Kommalapati, R., & Han, S.-E. (2017). Fluid-structure interaction analysis of NREL phase VI wind turbine: Aerodynamic force evaluation and structural analysis using FSI analysis. *Renewable Energy, 113*, 512–531.
6. Wang, T. (2012). A brief review on wind turbine aerodynamics. *Theoretical and Applied Mechanics Letters, 2*(6), 062001.
7. Herbert, G. J., Iniyan, S., Sreevalsan, E., & Rajapandian, S. (2007). A review of wind energy technologies. *Renewable and Sustainable Energy Reviews, 11*(6), 1117–1145.
8. Monroy, A., & Alvarez-Icaza, L. (2006). Real-time identification of wind turbine rotor power coefficient. In *Proceedings of the 45th IEEE Conference on Decision and Control* (pp. 3690–3695). New York: IEEE.
9. Xia, Y., Ahmed, K. H., & Williams, B. W. (2012). Wind turbine power coefficient analysis of a new maximum power point tracking technique. *IEEE Transactions on Industrial Electronics, 60*(3), 1122–1132.
10. Ragheb, M., & Ragheb, A. M. (2011). Wind turbines theory-the betz equation and optimal rotor tip speed ratio. *Fundamental and Advanced Topics in Wind Power, 1*(1), 19–38.

11. Çetin, N., Yurdusev, M., Ata, R., & Özdamar, A. (2005). Assessment of optimum tip speed ratio of wind turbines. *Mathematical and Computational Applications, 10*(1), 147–154.
12. Zhang, J., Cheng, M., Chen, Z., & Fu, X. (2008). Pitch angle control for variable speed wind turbines. In *2008 Third International Conference on Electric Utility Deregulation and Restructuring and Power Technologies* (pp. 2691–2696). New York: IEEE.
13. Mok, K. (2005). Identification of the power coefficient of wind turbines. In *IEEE Power Engineering Society General Meeting, 2005* (pp. 2078–2082). New York: IEEE.
14. Singh, M., & Santoso, S. (2011). Dynamic models for wind turbines and wind power plants. In *National Renewable Energy Lab.(NREL), Golden, CO (United States)*. Technical report.
15. Fernandez, L. M., Saenz, J. R., & Jurado, F. (2006). Dynamic models of wind farms with fixed speed wind turbines. *Renewable Energy, 31*(8), 1203–1230.
16. Muller, S., Deicke, M., & De Doncker, R. W. (2002). Doubly fed induction generator systems for wind turbines. *IEEE Industry Applications Magazine, 8*(3), 26–33.
17. Ekanayake, J. B., Holdsworth, L., Wu, X., & Jenkins, N. (2003). Dynamic modeling of doubly fed induction generator wind turbines. *IEEE Transactions on Power Systems, 18*(2), 803–809.
18. Bolik, S. M. (2004). *Modelling and analysis of variable speed wind turbines with induction generator during grid fault*. New York: Institut for Energiteknik, Aalborg Universitet.
19. Hansen, A. D., Soerensen, P., Iov, F., & Blaabjerg, F. (2004). Overall control strategy of variable speed doubly-fed induction generator wind turbine. In *Proceedings of Wind Power Nordic Conference*. New York: Chalmers Tekniska högskola.
20. Rodrigues, F. G. (2012). On equivalent expressions for the faraday's law of induction. *Revista Brasileira de Ensino de Fisica, 34*(1), 1–6.
21. Kundur, P. S., Balu, N. J., & Lauby, M. G. (2017). *Power system dynamics and stability*. Boca Raton: CRC Press.
22. Ekanayake, J., Holdsworth, L., & Jenkins, N. (2003). Control of DFIG wind turbines. *Power Engineer, 17*(1), 28–32.
23. Petersson, A., Thiringer, T., Harnefors, L., & Petru, T. (2005). Modeling and experimental verification of grid interaction of a DFIG wind turbine. *IEEE Transactions on Energy Conversion, 20*(4), 878–886.
24. Conroy, J., & Watson, R. (2007). Low-voltage ride-through of a full converter wind turbine with permanent magnet generator. *IET Renewable Power Generation, 1*(3), 182–189.
25. Conroy, J., & Watson, R. (2009). Aggregate modelling of wind farms containing full-converter wind turbine generators with permanent magnet synchronous machines: Transient stability studies. *IET Renewable Power Generation, 3*(1), 39–52.

# Chapter 6
# Power Electronics Interfaces and Controls

## 6.1 Modeling of Inverter

Some of the DERs generate DC power, e.g., PV, battery, and supercapacitor. Others generate very high or low frequency AC power, e.g., micro-turbine and wind. In order to integrate those DERs into the system, power-electronics interfaces are usually adopted such as inverter [1–3] and chopper [4–7]. Inverter is used to convert the DC voltage to AC voltage [8]. There are inverters for single phase [9] and three phases [10].

### 6.1.1 Single-Phase Inverter

The topology of a single-phase inverter is depicted in Fig. 6.1. Insulated Gate Bipolar Transistor (IGBT) is used for opening and closing the circuit super-fast (micro-second to nano-second) with high efficiency. We open and close IGBT in pairs to control the path electricity for generating AC voltage from DC source.

To converter DC voltage to AC voltage, the timing of opening and closing IGBT is very important [11]. Pulse Width Modulation (PWM) or Pulse Duration Modulation (PDM) is a typical strategy to control IGBT [12–14]. PWM is particularly suited for running inertial loads such as motors. They are not easily affected by the discrete switching of IGBT, because their inertia make them react slowly. Additionally, the PWM switching frequency must be high enough to smooth the waveform generated at the AC side.

Mathematically, the essential idea of PWM strategy is to determine the output voltage through comparing the carrier signal (the blue curve) and modulation signal (the red curve), as illustrated in Fig. 6.2 [15, 16]. The modulation signal can be adjusted by the controller of the inverter, which is introduced later.

© Springer Nature Switzerland AG 2022
Y. Li, *Cyber-Physical Microgrids*, https://doi.org/10.1007/978-3-030-80724-5_6

**Fig. 6.1** Topology of
single-phase inverter

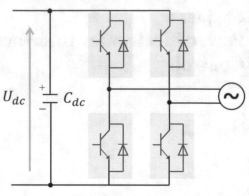

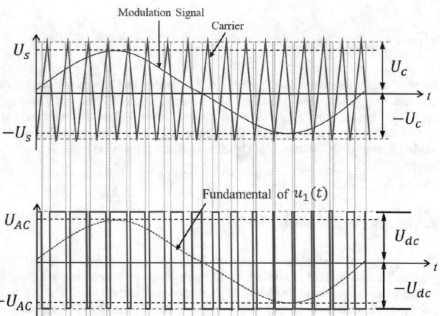

**Fig. 6.2** Carrier signal and modulation signal for PWM control

- When the modulation signal's value is larger than the carrier signal, the output
  voltage is $U_{dc}$, where $U_{dc}$ is the DC voltage.
- When the modulation signal's value is smaller than the carrier signal, the output
  voltage is $-U_{dc}$.

After getting the output voltage $U_{dc}$ or $-U_{dc}$, we can further express the output
voltage in the Fourier series. Further study shows the Root-Mean-Square (RMS)
value of the output voltage, $U_1$, is highly related to the DC voltage and the
modulation index $M$. Their relationship is illustrated in Fig. 6.3. The inverter is
highly recommended to operate in the linear region, as highlighted in Fig. 6.3, since

**Fig. 6.3** Relation between modulation index and output RMS voltage of the single phase inverter

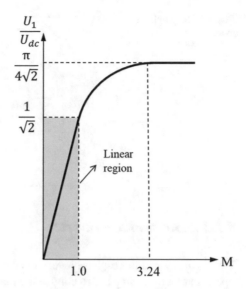

the output voltage can be simply controlled by adjusting the modulation index $M$. Note the RMS value is calculated below.

$$U_{RMS} = \sqrt{\frac{1}{T} \int_0^T u^2(t)dt,}$$ (6.1)

where,

- $U_{RMS}$ is the RMS value.
- $T$ is the period of the signal.
- $u(t)$ is the instantaneous signal.

When we use a filter to remove the high frequency harmonics of the output voltage, we can eventually get the nominal frequency AC voltage, which is expressed in (6.2).

$$u_1(t) = MU_{dc} \sin(\omega_s t - \phi),$$ (6.2)

where,

- $u_1(t)$ is the inverter's nominal frequency output voltage.
- $M$ is the modulation index.
- $\omega_s$ is the frequency of the modulation signal.
- $\phi$ is the initial angle of the modulation signal.

From (6.2), we can see the RMS value of the voltage $u_1(t)$ is $U_1 = MU_{dc}/\sqrt{2}$. Then, the ratio of $U_1/U_{dc}$ should be equal to $M/\sqrt{2}$, which verifies the result shown in Fig. 6.3.

**Fig. 6.4** Three-phase
inverter

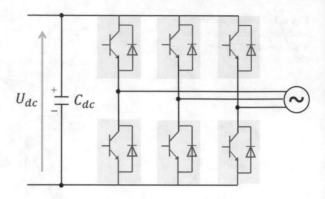

### 6.1.2 Three-Phase Inverter

The topology of a three-phase inverter is depicted in Fig. 6.4. IGBT devices are also
opened and closed in pairs to control the path electricity for generating three-phase
AC voltage from DC source. Similar to the analysis of the single-phase inverter,
when we use PWM technique, the RMS value of the line-to-line output voltage,
$U_{L1}$, is also highly related to the DC voltage and the modulation index $M$. Their
relationship is illustrated in Fig. 6.5. The inverter is usually operated in the linear
region. When we use a filter to remove the high frequency harmonics of the output
voltage, we can obtain the nominal frequency AC voltage. The line-to-line output
voltage $u_{LAB}(t)$ is given in (6.3) as an example.

$$
\begin{aligned}
u_{LAB1}(t) &= u_{LA1}(t) - u_{LB1}(t) \\
&= MU_{dc} \sin(\omega_s t - \phi) - MU_{dc} \sin\left(\omega_s t - \phi - \frac{2\pi}{3}\right) \\
&= \frac{\sqrt{3}MU_{dc}}{2} \cos\left(\omega_s t - \phi - \frac{\pi}{3}\right),
\end{aligned}
\tag{6.3}
$$

where,

- $u_{LAB1}(t)$ is the inverter's nominal frequency line-to-line output voltage.
- $M$ is the modulation index of the three-phase inverter.

From (6.3), we can see the RMS value of the voltage $u_{LAB1}(t)$ can be expressed
as follows,

$$
U_{L1} = \frac{\sqrt{3}MU_{dc}}{2\sqrt{2}}.
\tag{6.4}
$$

Then, the ration of $U_{L1}/U_{dc}$ should be equal to $\frac{\sqrt{3}M}{2\sqrt{2}}$, which verifies the result
shown in Fig. 6.5.

**Fig. 6.5** Relation between modulation index and output RMS voltage of the three-phase inverter

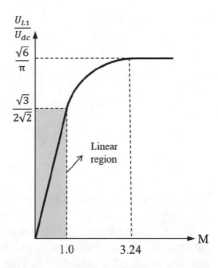

**Example 6.1**

A DC voltage source is connected to a three-phase inverter. The DC voltage is 1000 V.

Questions: What are the RMS values of the AC output voltages in the following two cases?

*Case 1*: The modulation index is 0.475.

*Case 2*: The modulation index is 0.95.

Solution: Based on (6.4), when the modulation index is 0.475, the RMS values can be calculated as below, respectively. ∎

$$U_{L1} = \frac{\sqrt{3} M U_{dc}}{2\sqrt{2}} = \frac{\sqrt{3}}{2\sqrt{2}} \times 0.475 \times 1000 = 290.88(V). \tag{6.5}$$

$$U_{L1} = \frac{\sqrt{3} M U_{dc}}{2\sqrt{2}} = \frac{\sqrt{3}}{2\sqrt{2}} \times 0.95 \times 1000 = 581.75(V). \tag{6.6}$$

## 6.2 Controller of Three-Phase Inverter

The modulation signal is playing an important role in generating AC output voltage [17, 18]. The function of inverter controller is to adjust the modulation index that corresponding to the modulation signals [19, 20]. As the output of three-phase inverter is three-phase AC signals and it is difficult to directly control AC signals, in

**Fig. 6.6** $dq0$ transformation

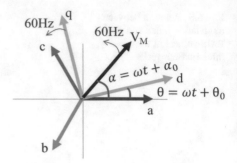

order to simplify the control strategy for AC signals, we first convert AC analysis to DC analysis.

Direct-quadrature-zero ($dq0$) transformation is adopted here [21, 22]. The essential idea of $dq0$ transformation is to rotate the reference frame of AC waveforms so that they become DC signals to simplify the analysis. Figure 6.6 illustrates the idea, where the reference frame of $dq0$ and voltage phasor $V$ are rotating at 60 Hz and the reference frame of $abc$ is stationary.

Mathematically, the conversion from $abc$ reference frame to $dq0$ frame can be expressed in (6.7).

$$\begin{bmatrix} v_d \\ v_q \\ v_0 \end{bmatrix} = k_1 \begin{bmatrix} \cos\theta & \cos(\theta - \frac{2\pi}{3}) & \cos(\theta + \frac{2\pi}{3}) \\ -\sin\theta & -\sin(\theta - \frac{2\pi}{3}) & -\sin(\theta + \frac{2\pi}{3}) \\ k_2 & k_2 & k_2 \end{bmatrix} \begin{bmatrix} v_a \\ v_b \\ v_c \end{bmatrix}, \tag{6.7}$$

where,

- $v_a$ is the instantaneous voltage of phase A, which can be expressed as $V_M \cos\alpha = V_M \cos(\omega t + \alpha_0)$.
- $v_b$ is the instantaneous voltage of phase B, which can be expressed as $V_M \cos(\alpha - \frac{2\pi}{3}) = V_M \cos(\omega t + \alpha_0 - \frac{2\pi}{3})$.
- $v_c$ is the instantaneous voltage of phase C, which can be expressed as $V_M \cos(\alpha + \frac{2\pi}{3}) = V_M \cos(\omega t + \alpha_0 + \frac{2\pi}{3})$.
- $\theta$ is the angle difference between the $d$-axis and $a$-axis, which can be expressed as $\theta = \omega t + \theta_0$.
- $k_1$ and $k_2$ are parameters to be determined.

Based on how we determine the values of $k_1$ and $k_2$, two transformations for the three-phase balanced system are introduced below.

**Constant Power $dq0$ Transformation** When we hope the power calculation format does not change before and after the transformation, $k_1$ is equal to $\sqrt{2/3}$ and $k_2$ is equal to $\sqrt{1/2}$. So that (6.7) can be rewritten as,

$$
\begin{bmatrix} v_d \\ v_q \\ v_0 \end{bmatrix} = \sqrt{\frac{2}{3}} \begin{bmatrix} \cos\theta & \cos(\theta - \frac{2\pi}{3}) & \cos(\theta + \frac{2\pi}{3}) \\ -\sin\theta & -\sin(\theta - \frac{2\pi}{3}) & -\sin(\theta + \frac{2\pi}{3}) \\ \sqrt{\frac{1}{2}} & \sqrt{\frac{1}{2}} & \sqrt{\frac{1}{2}} \end{bmatrix} \begin{bmatrix} v_a \\ v_b \\ v_c \end{bmatrix}. \tag{6.8}
$$

where, we define the transformation matrix $P_\theta$ as,

$$
P_\theta = \sqrt{\frac{2}{3}} \begin{bmatrix} \cos\theta & \cos(\theta - \frac{2\pi}{3}) & \cos(\theta + \frac{2\pi}{3}) \\ -\sin\theta & -\sin(\theta - \frac{2\pi}{3}) & -\sin(\theta + \frac{2\pi}{3}) \\ \sqrt{\frac{1}{2}} & \sqrt{\frac{1}{2}} & \sqrt{\frac{1}{2}} \end{bmatrix}. \tag{6.9}
$$

Then, we can get the inverse of $P_\theta$, as given in (6.10).

$$
P_\theta^{-1} = \sqrt{\frac{2}{3}} \begin{bmatrix} \cos\theta & -\sin\theta & \sqrt{\frac{1}{3}} \\ \cos(\theta - \frac{2\pi}{3}) & -\sin(\theta - \frac{2\pi}{3}) & \sqrt{\frac{1}{3}} \\ \cos(\theta + \frac{2\pi}{3}) & -\sin(\theta + \frac{2\pi}{3}) & \sqrt{\frac{1}{3}} \end{bmatrix} \tag{6.10}
$$

From (6.8), we can see $v_d$, $v_q$, and $v_0$ can be eventually calculated below.

$$
\begin{aligned}
v_d &= \sqrt{\frac{2}{3}}\Big(V_M \cos\theta \cos\alpha + V_M \cos\Big(\theta - \frac{2\pi}{3}\Big)\cos\Big(\alpha - \frac{2\pi}{3}\Big) \\
&\quad + V_M \cos\Big(\theta + \frac{2\pi}{3}\Big)\cos\Big(\alpha + \frac{2\pi}{3}\Big)\Big) \\
&= \sqrt{\frac{3}{2}} V_M \cos(\theta - \alpha)
\end{aligned} \tag{6.11}
$$

$$
\begin{aligned}
v_q &= \sqrt{\frac{2}{3}}\Big(-V_M \sin\theta \cos\alpha - V_M \sin\Big(\theta - \frac{2\pi}{3}\Big)\cos\Big(\alpha - \frac{2\pi}{3}\Big) \\
&\quad - V_M \sin\Big(\theta + \frac{2\pi}{3}\Big)\cos\Big(\alpha + \frac{2\pi}{3}\Big)\Big) \\
&= -\sqrt{\frac{3}{2}} V_M \sin(\theta - \alpha)
\end{aligned} \tag{6.12}
$$

$$
v_0 = \sqrt{\frac{2}{3}}\Big(\sqrt{\frac{1}{2}}V_M \cos\alpha + \sqrt{\frac{1}{2}}V_M \cos\Big(\alpha - \frac{2\pi}{3}\Big) + \sqrt{\frac{1}{2}}V_M \cos\Big(\alpha + \frac{2\pi}{3}\Big)\Big) = 0 \tag{6.13}
$$

The three-phase complex power is calculated in (6.14).

$$
S = v_a i_a^* + v_b i_b^* + v_c i_c^* \tag{6.14}
$$

where, * means the conjugate value.

Based on (6.14) and considering $v_{abc} = P_\theta^{-1} v_{dq0}$ and $i_{abc} = P_\theta^{-1} i_{dq0}$, we can get the complex power calculation in the constant power dq0 transformation, as derived in (6.15).

$$
\begin{aligned}
S = v_a i_a^* + v_b i_b^* + v_c i_c^* &= (P_\theta^{-1} v_{dq0})^T P_\theta^{-1} i_{dq0} \\
&= (v_{dq0})^T (P_\theta^{-1})^T P_\theta^{-1} i_{dq0} \\
&= (v_{dq0})^T i_{dq0} \\
&= v_d i_d + v_q i_q + v_0 i_0.
\end{aligned}
\tag{6.15}
$$

where, $(P_\theta^{-1})^T P_\theta^{-1} = 1$.

From (6.11) to (6.15), we can see in the constant power dq0 transformation, the $dq0$ frame complex power calculation can keep the same format, but the $dq0$ frame voltage (or current) amplitude is different from the $abc$ frame calculation.

**Constant Amplitude $dq0$ Transformation**   When we hope the voltage and current calculation format does not change before and after the transformation, $k_1$ is equal to 2/3 and $k_2$ is equal to 1/2. So that (6.7) can be rewritten as,

$$
\begin{bmatrix} v_d \\ v_q \\ v_0 \end{bmatrix} = \frac{2}{3} \begin{bmatrix} \cos\theta & \cos(\theta - \frac{2\pi}{3}) & \cos(\theta + \frac{2\pi}{3}) \\ -\sin\theta & -\sin(\theta - \frac{2\pi}{3}) & -\sin(\theta + \frac{2\pi}{3}) \\ \frac{1}{2} & \frac{1}{2} & \frac{1}{2} \end{bmatrix} \begin{bmatrix} v_a \\ v_b \\ v_c \end{bmatrix}.
\tag{6.16}
$$

where, we define the transformation matrix $P_\theta$ as,

$$
P_\theta = \frac{2}{3} \begin{bmatrix} \cos\theta & \cos(\theta - \frac{2\pi}{3}) & \cos(\theta + \frac{2\pi}{3}) \\ -\sin\theta & -\sin(\theta - \frac{2\pi}{3}) & -\sin(\theta + \frac{2\pi}{3}) \\ \frac{1}{2} & \frac{1}{2} & \frac{1}{2} \end{bmatrix}
\tag{6.17}
$$

Then, we can get the inverse of $P_\theta$,

$$
P_\theta^{-1} = \begin{bmatrix} \cos\theta & -\sin\theta & 1 \\ \cos(\theta - \frac{2\pi}{3}) & -\sin(\theta - \frac{2\pi}{3}) & 1 \\ \cos(\theta + \frac{2\pi}{3}) & -\sin(\theta + \frac{2\pi}{3}) & 1 \end{bmatrix}
\tag{6.18}
$$

From (6.16), we can see $v_d$, $v_q$, and $v_0$ can be eventually calculated below.

$$
\begin{aligned}
v_d &= \frac{2}{3}\Big(V_M \cos\theta \cos\alpha + V_M \cos(\theta - \frac{2\pi}{3}) \cos(\alpha - \frac{2\pi}{3}) \\
&\quad + V_M \cos(\theta + \frac{2\pi}{3}) \cos(\alpha + \frac{2\pi}{3})\Big) \\
&= V_M \cos(\theta - \alpha)
\end{aligned}
\tag{6.19}
$$

$$v_q = \frac{2}{3}\left(-V_M \sin\theta \cos\alpha - V_M \sin\left(\theta - \frac{2\pi}{3}\right)\cos\left(\alpha - \frac{2\pi}{3}\right)\right.$$
$$\left. - V_M \sin\left(\theta + \frac{2\pi}{3}\right)\cos\left(\alpha + \frac{2\pi}{3}\right)\right) \tag{6.20}$$
$$= -V_M \sin(\theta - \alpha)$$

$$v_0 = \frac{2}{3}\left(\frac{1}{2}V_M \cos\alpha + \frac{1}{2}V_M \cos\left(\alpha - \frac{2\pi}{3}\right) + \frac{1}{2}V_M \cos\left(\alpha + \frac{2\pi}{3}\right)\right) = 0 \tag{6.21}$$

When we convert current's $abc$ signals to $dq0$ signals, we can also get a similar format. Because the constant amplitude $dq0$ transformation simplifies the conversion of voltage and current signals, this transformation is used as an example when we introduce the inverter control system.

Considering (6.14), $v_{abc} = P_\theta^{-1} v_{dq0}$ and $i_{abc} = P_\theta^{-1} i_{dq0}$, we can get the complex power calculation in the constant amplitude $dq0$ transformation, as shown in (6.22).

$$S = v_a i_a^* + v_b i_b^* + v_c i_c^* = (P_\theta^{-1} v_{dq0})^T P_\theta^{-1} i_{dq0}$$
$$= (v_{dq0})^T (P_\theta^{-1})^T P_\theta^{-1} i_{dq0}$$
$$= \frac{3}{2}(v_{dq0})^T i_{dq0} \tag{6.22}$$
$$= \frac{3}{2}(v_d i_d + v_q i_q + v_0 i_0).$$

where $(P_\theta^{-1})^T P_\theta^{-1} = \frac{3}{2}$.

From (6.19) to (6.22), we can see in the constant amplitude $dq0$ transformation, the voltage (or current) amplitude can keep same, but the $dq0$ frame complex power calculation is different from the $abc$ frame complex power calculation.

From (6.19) and (6.20), we can see when $\theta$ is equal to the phase A's angle $\alpha$, we can further simply the transformation because in this case we have $v_d = V_M$ and $v_q = 0$. Hence, by only controlling the $d$-axis signal, we can adjust the voltage amplitude $V_M$.

A typical system integration using inverter is given in Fig. 6.7. It includes Phase Lock Loop (PLL), Double Loop Controller (Outer Controller and Inner Controller), and PWM Generator.

## 6.2.1 Phase Lock Loop (PLL)

As introduced, when $\theta$ is equal to the phase A's angle $\alpha$ as shown in Fig. 6.6, we can have $v_d = V_M$ and $v_q = 0$. Therefore, when $v_q = 0$, we can guarantee $\theta = \alpha$.

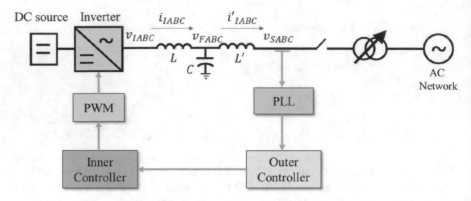

**Fig. 6.7** System integration using inverter

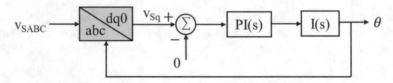

**Fig. 6.8** Example of phase lock loop

**Fig. 6.9** $dq0$ transformation
when the $d$-axis is in phase
with the voltage phasor

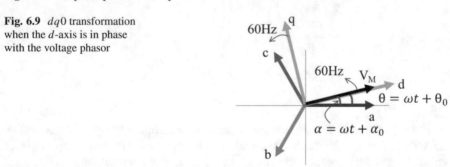

The function of PLL is to have $v_q = 0$, so that we can have the $d$-axis in phase with the voltage phasor, i.e., $\theta = \alpha$. A PLL example of realizing the above function is shown in Fig. 6.8, where PI(s) is a proportional-integral controller and I(s) is an integral controller. Therefore, when the phase A's angle is locked, Fig. 6.6 will become Fig. 6.9, which shows the $d$-axis in phase with the voltage phasor.

*Note 6.1*  when the phase A's angle is locked, $\theta$ is changing over time because the $d$-axis and the voltage phasor are still rotating.

### 6.2.2 Double Loop Controller

The double loop controller is the essential part of the inverter's control system. In order to understand its function, we need to look at the output circuit in Fig. 6.7. Applying KVL to the inductance $L$, we can get the following expression which is in the $abc$ format.

$$\begin{cases} L\dfrac{di_{IA}}{dt} = v_{IA} - v_{FA} \\[2mm] L\dfrac{di_{IB}}{dt} = v_{IB} - v_{FB} \\[2mm] L\dfrac{di_{IC}}{dt} = v_{IC} - v_{FC}, \end{cases} \tag{6.23}$$

where,

- $v_{IABC} = [v_{IA}, v_{IB}, v_{IC}]$ is the three-phase instantaneous phase to ground voltage of the inverter's output voltage.
- $i_{IABC} = [i_{IA}, i_{IB}, i_{IC}]$ is the three-phase instantaneous current through $L$.
- $v_{FABC} = [v_{FA}, v_{FB}, v_{FC}]$ is the three-phase instantaneous phase to ground voltage of the capacitor.

Applying KVL to the inductance $L'$, we can get (6.24) which is also in the $abc$ format.

$$\begin{cases} L'\dfrac{di'_{IA}}{dt} = v_{FA} - v_{SA} \\[2mm] L'\dfrac{di'_{IB}}{dt} = v_{FB} - v_{SB} \\[2mm] L'\dfrac{di'_{IC}}{dt} = v_{FC} - v_{SC}, \end{cases} \tag{6.24}$$

where,

- $i'_{IABC} = [i'_{IA}, i'_{IB}, i'_{IC}]$ is the three-phase instantaneous current through $L'$.
- $v_{SABC} = [v_{SA}, v_{SB}, v_{SC}]$ is the three-phase instantaneous phase to ground voltage of the AC network.

Correspondingly, the current through the capacitor $C$ can be calculated through (6.25), where KCL is applied.

$$
\begin{cases}
C\dfrac{dv_{FA}}{dt} = i_{IA} - \dfrac{1}{L'}\displaystyle\int (v_{FA} - v_{SA})dt \\[2ex]
C\dfrac{dv_{FB}}{dt} = i_{IB} - \dfrac{1}{L'}\displaystyle\int (v_{FB} - v_{SB})dt \\[2ex]
C\dfrac{dv_{FC}}{dt} = i_{IC} - \dfrac{1}{L'}\displaystyle\int (v_{FC} - v_{SC})dt.
\end{cases}
\tag{6.25}
$$

In practice, the capacitor $C$ is usually very small, so (6.25) can be ignored. Then (6.23) and (6.24) can be further simplified as shown in (6.26), where $\hat{L} = L + L'$.

$$
\begin{cases}
\hat{L}\dfrac{di_{IA}}{dt} = v_{IA} - v_{SA} \\[2ex]
\hat{L}\dfrac{di_{IB}}{dt} = v_{IB} - v_{SB} \\[2ex]
\hat{L}\dfrac{di_{IC}}{dt} = v_{IC} - v_{SC}.
\end{cases}
\tag{6.26}
$$

When we transform (6.26) into the $dq0$ reference frame, we can get the following expression of $dq$ current.

$$
\begin{cases}
\hat{L}\dfrac{di_{Id}}{dt} = v_{Id} - v_{Sd} - \omega\hat{L}i_{Iq} \\[2ex]
\hat{L}\dfrac{di_{Iq}}{dt} = v_{Iq} - v_{Sq} + \omega\hat{L}i_{Id},
\end{cases}
\tag{6.27}
$$

where,

- $[i_{Id}, i_{Iq}]$ is $dq$ current obtained from the $dq0$ transformation.
- $[v_{Id}, v_{Iq}]$ and $[v_{Sd}, v_{Sq}]$ are $dq$ voltages obtained from the $dq0$ transformation.

Based on (6.27), we can rewrite it to get the output voltage of inverter, as shown in (6.28).

$$
\begin{cases}
v_{Id} = \hat{L}\dfrac{di_{Id}}{dt} + v_{Sd} + \omega\hat{L}i_{Iq} \\[2ex]
v_{Iq} = \hat{L}\dfrac{di_{Iq}}{dt} + v_{Sq} - \omega\hat{L}i_{Id}.
\end{cases}
\tag{6.28}
$$

From (6.28), we can see if the output voltage $[v_{Id}, v_{Iq}]$ can be adjusted through controlling the current $[i_{Id}, i_{Iq}]$. It describes the function of the inner controller. An inner controller example is shown in Fig. 6.10, where a proportional-integral controller is used to control the current.

For the outer controller, it could control the voltage, active power, or reactive power. Examples will be introduced in the integration of DERs.

**Fig. 6.10** Example of inner controller for inverter

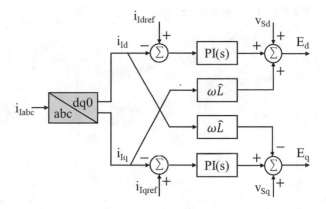

### 6.2.3 PWM Generator

As we can see from Fig. 6.10, its outputs $[E_d, E_q]$ are DC signals; however, AC modulation signal is needed in Fig. 6.2. So, the PWM generator has two functions, as introduced below.

- Convert the DC control signal to the *abc* reference frame through an inverse $dq0$ transformation.
- Compare the carrier signal with the modulation signal, as shown in Fig. 6.2, to generate the switching signals for controlling IGBT.

## 6.3 Chopper

A chopper is a static power electronics device that converts DC input to a variable DC voltage/power. It is a DC to DC converter. Based on the relation between input DC voltage and output DC voltage, there are three major types of choppers [23, 24], i.e., boost chopper, buck chopper, and buck-boost chopper.

### 6.3.1 Boost Chopper

The function of boost chopper is to convert a low input voltage to a higher voltage. An example of boost chopper is given in Fig. 6.11.

To analyze the operation of the boost chopper, duty ratio is introduced in Fig. 6.12 and defined in (6.29), from which we can see $0 \leq D \leq 1$.

$$D = \frac{T_{on}}{T_{on} + T_{off}},$$ (6.29)

**Fig. 6.11** Example of boost chopper

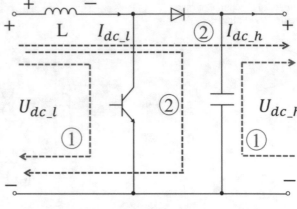

**Fig. 6.12** Duty ratio

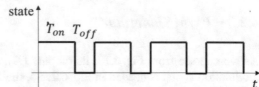

where,

- $T_{on}$ is time period during which the IGBT is closed. It is called *On State*.
- $T_{off}$ is time period during which the IGBT is open. It is called *Off State*.

Assume Fig. 6.11 shows an ideal boost chopper, which means the power loss is ignored.

*On State*  When the IGBT is closed, the DC current will go through ①. That is no current through the diode. Analyzing the current through the inductor $L$ and voltage over the capacitor $C$, we can get the following equations.

$$\begin{cases} L\dfrac{dI_{dc\_l}}{dt} = U_{dc\_l} = u_{L(on)} \\ C\dfrac{dU_{dc\_h}}{dt} = I_{dc\_h} = i_{C(on)}. \end{cases} \tag{6.30}$$

*Off State*  When the IGBT is open, the DC current will go through ②. That is no current through the IGBT. Analyzing the current through the inductor $L$ and voltage over the capacitor $C$, we can get the following equations.

$$\begin{cases} L\dfrac{dI_{dc\_l}}{dt} = U_{dc\_l} - U_{dc\_h} = u_{L(off)} \\ C\dfrac{dU_{dc\_h}}{dt} = I_{dc\_h} - I_{dc\_l} = i_{C(off)}. \end{cases} \tag{6.31}$$

From the above analysis, we can see as the IGBT changes its state (on or off), the inductor will charge or discharge power. The average of inductor's voltage over one period is expected to be zero, as discussed below.

For one thing, the average inductor's voltage is calculated in (6.32), where $T_s$ is the period and $u_L(t) = L\frac{dI_{dc\_l}}{dt}$.

$$\bar{u}_L = \frac{1}{T_s}\int_t^{t+T_s} u_L(t)dt = \frac{1}{T_s}\int_t^{t+T_s} L\frac{dI_{dc\_l}}{dt}dt = \frac{L}{T_s}\int_t^{t+T_s} dI_{dc\_l}. \qquad (6.32)$$

During the IGBT changes its state (on or off), the inductor charges or discharges power. Considering the energy stored in the inductor $E = \frac{1}{2}LI_{dc\_l}^2$, within one period, the change of energy should be equal to zero. So we can have the following expression.

$$dE = LI_{dc\_l}dI_{dc\_l} = 0. \qquad (6.33)$$

Hence, $dI_{dc\_l} = 0$. Substitute $dI_{dc\_l} = 0$ into (6.32), we have $\bar{u}_L = 0$.

For another, the average inductor's voltage can also be calculated in (6.34).

$$\begin{aligned} \bar{u}_L &= \frac{1}{T_s}\int_t^{t+T_s} u_L(t)dt \\ &= \frac{1}{T_s}\left(\int_t^{t+DT_s} u_{L(on)}(t)dt + \int_{t+DT_s}^{t+T_s} u_{L(off)}(t)dt\right). \end{aligned} \qquad (6.34)$$

Substitute (6.30) and (6.31) into (6.34), we can then get (6.35).

$$\begin{aligned} \bar{u}_L &= \frac{1}{T_s}\left(D \cdot T_s \cdot U_{dc\_l} + (1-D) \cdot T_s \cdot (U_{dc\_l} - U_{dc\_h})\right) \\ &= D \cdot U_{dc\_l} + (1-D) \cdot (U_{dc\_l} - U_{dc\_h}). \end{aligned} \qquad (6.35)$$

As analyzed above, (6.35) is expected to be zero; and thus, we can have the following voltage expression.

$$U_{dc\_h} = \frac{1}{1-D}U_{dc\_l}. \qquad (6.36)$$

Because $0 \leq D \leq 1$, the output voltage $U_{dc\_h}$ is larger than the input voltage $U_{dc\_l}$, which verifies it is a boost chopper. That is the boost chopper converts a low input voltage to a higher voltage.

From the above analysis, we can also see as the IGBT changes its state, the capacitor will also charge or discharge power. The average of capacitor's current over one period, as calculated in (6.37), is expected to be zero, as discussed below.

For one thing, the average capacitor's current is calculated in (6.37), where $T_s$ is the period and $i_C(t) = C \frac{dU_{dc\_h}}{dt}$.

$$\bar{i}_C = \frac{1}{T_s} \int_t^{t+T_s} i_C(t) dt = \frac{1}{T_s} \int_t^{t+T_s} C \frac{dU_{dc\_h}}{dt} dt = \frac{L}{T_s} \int_t^{t+T_s} dU_{dc\_h}. \qquad (6.37)$$

During the IGBT changes its state (on or off), the capacitor charges or discharges power. Considering the energy stored in the capacitor $E = \frac{1}{2} C U_{dc\_h}^2$, within one period, the change of energy should be equal to zero. So we can have the following expression.

$$dE = C U_{dc\_h} dU_{dc\_h} = 0. \qquad (6.38)$$

Hence, $dU_{dc\_h} = 0$. Substitute $ddU_{dc\_h}$ into (6.37), we have $\bar{i}_C = 0$.

For another, the average capacitor's current can also be calculated in (6.39).

$$\bar{i}_C = \frac{1}{T_s} \int_t^{t+T_s} i_C(t) dt = \frac{1}{T_s} \left( \int_t^{t+DT_s} i_{C(on)}(t) dt + \int_{t+DT_s}^{t+T_s} i_{C(off)}(t) dt \right)$$

$$= \frac{1}{T_s} \left( D \cdot T_s \cdot I_{dc\_l} + (1 - D) \cdot T_s \cdot (I_{dc\_h} - I_{dc\_l}) \right)$$

$$= D \cdot I_{dc\_h} + (1 - D) \cdot (I_{dc\_h} - I_{dc\_l}).$$
$$(6.39)$$

When we have (6.39) equal to zero, we can get the following current expression.

$$I_{dc\_h} = (1 - D) I_{dc\_l}. \qquad (6.40)$$

Based on (6.36) and (6.40), the output active power can be calculated in (6.41), which verifies the power loss in the ideal boost chopper is zero.

$$P_h = U_{dc\_h} \cdot I_{dc\_h} = \frac{1}{1 - D} U_{dc\_l} \cdot (1 - D) I_{dc\_l} = U_{dc\_l} \cdot I_{dc\_l} = P_l. \qquad (6.41)$$

### 6.3.2   Buck Chopper

The function of buck chopper is to convert a high input voltage to a lower voltage. An example of buck chopper is given in Fig. 6.13.

We also assume Fig. 6.13 shows an ideal buck chopper, which means the power loss is ignored. Then we can analyze the on state and off state as below.

*On State*  When the IGBT is closed, the DC current will go through ①. That is the capacitor discharges power and no current goes through the diode. Analyzing the current through the inductor $L$ and voltage over the capacitor $C$, we can get the

**Fig. 6.13** Example of buck chopper

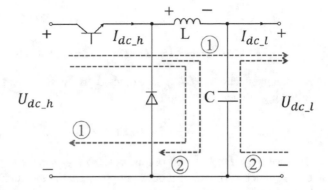

following equations.

$$
\begin{cases}
L\dfrac{dI_{dc\_h}}{dt} = U_{dc\_h} - U_{dc\_l} = u_{L(on)} \\[2mm]
C\dfrac{dU_{dc\_l}}{dt} = I_{dc\_h} - I_{dc\_l} = i_{C(on)}.
\end{cases}
\tag{6.42}
$$

*Off State* When the IGBT is open, the DC current will go through ②. That is the capacitor charges power and no current goes through the IGBT. Analyzing the current through the inductor $L$ and voltage over the capacitor $C$, we can get the following equations.

$$
\begin{cases}
L\dfrac{dI_{dc\_h}}{dt} = -U_{dc\_l} = u_{L(off)} \\[2mm]
C\dfrac{dU_{dc\_l}}{dt} = I_{dc\_h} = i_{C(off)}.
\end{cases}
\tag{6.43}
$$

From the above analysis, we can see as the IGBT changes its state (on or off), the inductor will charge and discharge power. The average of inductor's voltage over one period is expected to be zero, as discussed in the boost chopper. The average inductor's voltage is calculated in (6.44), where $T_s$ is the period.

$$
\begin{aligned}
\bar{u}_L &= \frac{1}{T_s}\int_t^{t+T_s} u_L(t)dt \\
&= \frac{1}{T_s}\Big(\int_t^{t+DT_s} u_{L(on)}(t)dt + \int_{t+DT_s}^{t+T_s} u_{L(off)}(t)dt\Big).
\end{aligned}
\tag{6.44}
$$

Substitute (6.42) and (6.43) into (6.44), we can then get (6.45).

$$\bar{u}_L = \frac{1}{T_s}\big(D \cdot T_s \cdot (U_{dc\_h} - U_{dc\_l}) + (1 - D) \cdot T_s \cdot (-U_{dc\_l})\big) \tag{6.45}$$
$$= D \cdot (U_{dc\_h} - U_{dc\_l}) + (1 - D) \cdot (-U_{dc\_l}).$$

As analyzed above, (6.45) is expected to be zero; and thus, we can have the following voltage expression.

$$U_{dc\_l} = D \cdot U_{dc\_h}. \tag{6.46}$$

Because $0 \leq D \leq 1$, the output voltage $U_{dc\_l}$ is smaller than the input voltage $U_{dc\_h}$, which verifies it is a buck chopper. That is the buck chopper converts a high input voltage to a lower voltage.

From the above analysis, we can also see as the IGBT changes its state, the capacitor will also charge and discharge power. The average of capacitor's current over one period, as calculated in (6.47), is expected to be zero.

$$\bar{i}_C = \frac{1}{T_s} \int_t^{t+T_s} i_C(t)dt = \frac{1}{T_s}\Big( \int_t^{t+DT_s} i_{C(on)}(t)dt + \int_{t+DT_s}^{t+T_s} i_{C(off)}(t)dt \Big)$$
$$= \frac{1}{T_s}\big(D \cdot T_s \cdot (I_{dc\_h} - I_{dc\_l}) + (1 - D) \cdot T_s \cdot I_{dc\_h}\big)$$
$$= D \cdot (I_{dc\_h} - I_{dc\_l}) + (1 - D) \cdot I_{dc\_h}. \tag{6.47}$$

When we have (6.47) equal to zero, we can get the following current expression.

$$I_{dc\_l} = \frac{1}{D} I_{dc\_h}. \tag{6.48}$$

Based on (6.46) and (6.48), the output active power can be calculated in (6.49), which verifies the power loss in the ideal buck chopper is zero.

$$P_l = U_{dc\_l} \cdot I_{dc\_l} = D \cdot U_{dc\_h} \cdot \frac{1}{D} I_{dc\_h} = U_{dc\_h} \cdot I_{dc\_h} = P_h. \tag{6.49}$$

### 6.3.3  Buck-Boost Chopper

The function of buck-boost chopper is to convert a DC input voltage into a different level output voltage. An example of buck-boost chopper is given in Fig. 6.14.

We also assume Fig. 6.14 shows an ideal buck-boost chopper, which means the power loss is ignored. Then we can analyze the on state and off state as below.

*On State*  When the IGBT is closed, the DC current will go through ①. That is the capacitor discharges power, no current goes through the diode, and the inductor

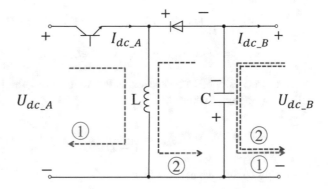

**Fig. 6.14** Example of buck-boost chopper

charges power. Analyzing the current through the inductor $L$ and voltage over the capacitor $C$, we can get the following equations.

$$\begin{cases} L\dfrac{dI_{dc\_A}}{dt} = U_{dc\_A} = u_{L(on)} \\ C\dfrac{dU_{dc\_B}}{dt} = -I_{dc\_B} = i_{C(on)}. \end{cases} \tag{6.50}$$

*Off State* When the IGBT is open, the DC current will go through ②. That is the capacitor charges power, no current goes through the IGBT, and the inductor discharges power. Analyzing the current through the inductor $L$ and voltage over the capacitor $C$, we can get the following equations.

$$\begin{cases} L\dfrac{dI_{dc\_A}}{dt} = -U_{dc\_B} = u_{L(off)} \\ C\dfrac{dU_{dc\_B}}{dt} = I_{dc\_A} = i_{C(off)}. \end{cases} \tag{6.51}$$

From the above analysis, we can see as the IGBT changes its state (on or off), the inductor will charge and discharge power. The average of inductor's voltage over one period is expected to be zero, as discussed in the boost chopper. The average inductor's voltage is calculated in (6.52).

$$\begin{aligned} \bar{u}_L &= \frac{1}{T_s}\int_t^{t+T_s} u_L(t)dt \\ &= \frac{1}{T_s}\Big(\int_t^{t+DT_s} u_{L(on)}(t)dt + \int_{t+DT_s}^{t+T_s} u_{L(off)}(t)dt\Big). \end{aligned} \tag{6.52}$$

Substitute (6.50) and (6.51) into (6.52), we can then get (6.53).

$$\bar{u}_L = \frac{1}{T_s}\big(D \cdot T_s \cdot U_{dc\_A} + (1 - D) \cdot T_s \cdot (-U_{dc\_B})\big)$$

$$= D \cdot U_{dc\_A} + (1 - D) \cdot (-U_{dc\_B}). \tag{6.53}$$

As analyzed above, (6.53) is expected to be zero; and thus, we can have the following voltage expression.

$$U_{dc\_B} = \frac{D}{1 - D} \cdot U_{dc\_A} = k \cdot U_{dc\_A}. \tag{6.54}$$

Analyzing (6.54), we can see that:

- When $1/2 < D < 1$, we have $k > 1$, which means the output voltage is larger than the input voltage. So, the chopper works as a boost converter.
- When $D = 1/2$, we have $k = 1$, which means the output voltage is equal to the input voltage.
- When $0 < D < 1/2$, we have $k < 1$, which means the output voltage is smaller than the input voltage. So, the chopper works as a buck converter.

From the above analysis, we can also see as the IGBT changes its state, the capacitor will also charge OR discharge power. The average of capacitor's current over one period, as calculated in (6.55), is expected to be zero.

$$\bar{i}_C = \frac{1}{T_s}\int_t^{t+T_s} i_C(t)dt = \frac{1}{T_s}\bigg(\int_t^{t+DT_s} i_{C(on)}(t)dt$$

$$+ \int_{t+DT_s}^{t+T_s} i_{C(off)}(t)dt\bigg) = \frac{1}{T_s}\big(D \cdot T_s \cdot (-I_{dc\_B}) + (1 - D) \cdot T_s \cdot I_{dc\_A}\big)$$

$$= D \cdot (-I_{dc\_B}) + (1 - D) \cdot I_{dc\_A}. \tag{6.55}$$

When we have (6.55) equal to zero, we can get the following current expression.

$$I_{dc\_B} = \frac{1 - D}{D} \cdot I_{dc\_A}. \tag{6.56}$$

Based on (6.54) and (6.56), the output active power can be calculated in (6.57), which verifies the power loss in the ideal buckboost chopper is zero.

$$P_B = U_{dc\_B} \cdot I_{dc\_B} = \frac{D}{1 - D} \cdot U_{dc\_A} \cdot \frac{1 - D}{D} \cdot I_{dc\_A} = U_{dc\_A} \cdot I_{dc\_A} = P_A. \tag{6.57}$$

**Example 6.2**
A buckboost chopper as shown in Fig. 6.14 is used to connect a DC voltage to a power load. The DC voltage is 10 V and current is 100 A.
Questions:
*Case 1*: When the duty ratio is equal to 0.333, what is the output voltage? What is the power load's DC current? What is the function of the buckboost chopper?
*Case 2*: When the duty ratio is equal to 0.375, what is the output voltage? What is the power load's DC current? What is the function of the buckboost chopper?
*Case 3*: When the duty ratio is equal to 0.5, what is the output voltage? What is the power load's DC current? What is the function of the buckboost chopper?
*Case 4*: When the duty ratio is equal to 0.6, what is the output voltage? What is the power load's DC current? What is the function of the buckboost chopper?

Solution:
*Case 1*: According to (6.54), when the duty ratio is equal to 0.333, we can have the following voltage calculation.

$$U_{dc\_B} = \frac{D}{1-D} \cdot U_{dc\_A} = \frac{0.333}{1-0.333} \times 10 = 5(\text{V}). \tag{6.58}$$

Correspondingly, according to (6.56), when the duty ratio is equal to 0.333, we can have the following current calculation.

$$I_{dc\_B} = \frac{1-D}{D} \cdot I_{dc\_A} = \frac{1-0.333}{0.333} \times 100 = 200(\text{A}). \tag{6.59}$$

From above calculation, we can verify the power generated by the DC voltage source is equal to the power used by the load. From the comparison of input voltage and output voltage, we can see the buckboost chopper is working in the buck mode.
Similar to Case 1, we can compute other cases as below.
*Case 2*:

$$U_{dc\_B} = \frac{D}{1-D} \cdot U_{dc\_A} = \frac{0.375}{1-0.375} \times 10 = 6(\text{V}). \tag{6.60}$$

$$I_{dc\_B} = \frac{1-D}{D} \cdot I_{dc\_A} = \frac{1-0.375}{0.375} \times 100 = 166.7(\text{A}). \tag{6.61}$$

From the comparison of input voltage and output voltage, we can see the buckboost chopper is working in the buck mode.
*Case 3*:

$$U_{dc\_B} = \frac{D}{1 - D} \cdot U_{dc\_A} = \frac{0.5}{1 - 0.5} \times 10 = 10 \text{(V)}. \tag{6.62}$$

$$I_{dc\_B} = \frac{1 - D}{D} \cdot I_{dc\_A} = \frac{1 - 0.5}{0.5} \times 100 = 100 \text{(A)}. \tag{6.63}$$

From the comparison of input voltage and output voltage, we can see the buckboost chopper does not change voltage or current.
*Case 4*:

$$U_{dc\_B} = \frac{D}{1 - D} \cdot U_{dc\_A} = \frac{0.6}{1 - 0.6} \times 10 = 15 \text{(V)}. \tag{6.64}$$

$$I_{dc\_B} = \frac{1 - D}{D} \cdot I_{dc\_A} = \frac{1 - 0.6}{0.6} \times 100 = 66.7 \text{(A)}. \tag{6.65}$$

From the comparison of input voltage and output voltage, we can see the buckboost chopper is working in the boost mode. ∎

## 6.4   System Integration of PV

PV can convert sunlight to electric energy. But it can only produce electricity during the day time. For instance, if we have roof top solar panels, we can use PV energy to support the electricity consumption of a community in the day time, but we will have no power at night. Meanwhile, PV panels have fluctuating power output due to the frequent changes of irradiance and temperature. To solve these issues, one tractable solution is to integrate PV panels to the power grids. Considering the DC power output of PV array, power electronic interface is used to converter DC output to AC power, then connect to the system. Depending on the topology of the power electronic interface, single stage or two-stage integration can be adopted based on the needs and PV output voltage [25, 26].

### 6.4.1   Single Stage Integration

In the single stage integration, an inverter is used to converter DC voltage to AC voltage. MPPT function is implemented in the inverter controller. This system integration has simple interface topology and high energy efficiency; and meanwhile, it can help reduce cost. However, because only inverter is used, its controller is relatively complex, so it is usually difficult to coordinate the control parameters. Figure 6.15 shows the topology of single stage system integration.

The controller of inverter is given in Fig. 6.7, where the outer controller is used to implement MPPT. To introduce the outer controller, we will start from the analysis

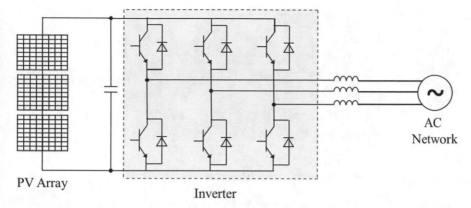

PV Array

Inverter

**Fig. 6.15** Topology of single stage PV system integration

**Fig. 6.16** Typical outer
controller for PV inverter in
the single stage integration

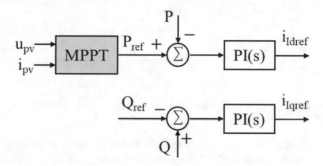

of power output. Based on the constant amplitude $dq0$ transformation, the active
power and reactive power can be calculated through (6.66) and (6.67), respectively.

$$P = \frac{3}{2}(v_{Sd}i_{Id} + v_{Sq}i_{Iq} + 2v_{S0}i_{I0}) \tag{6.66}$$

$$Q = \frac{3}{2}(v_{Sq}i_{Id} - v_{Sd}i_{Iq}) \tag{6.67}$$

Considering the function of PLL is to have $v_{Sq} = 0$, we can see the active power
$P$ is highly related to $i_{Id}$ and the reactive power $Q$ is related to $i_{Iq}$. So that in the
outer controller, we can use active power reference to generate the reference signal
$i_{Idref}$ in Fig. 6.10 and use reactive power reference to generate the reference signal
$i_{Iqref}$. A typical outer controller is given in Fig. 6.16.

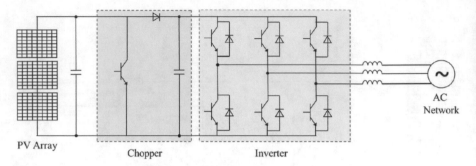

**Fig. 6.17**  Topology of two-stage PV system integration

**Table 6.1**  Changes of duty ratio in the P&O MPPT method

|   | Voltage perturbation | Active power change | Duty ratio change |
|---|---|---|---|
| ① | Positive | Positive | Negative |
| ② | Positive | Negative | Positive |
| ③ | Negative | Positive | Positive |
| ④ | Negative | Negative | Negative |

## 6.4.2  Two-Stage Integration

In the two-stage integration, a boost chopper is usually used to increase the DC output voltage of PV array and then an inverter is used to converter DC voltage to AC voltage. MPPT function is typically implemented in the boost chopper's controller. Figure 6.17 shows the topology of two-stage system integration.

In the two-stage integration, the function of the inverter's controller is usually to maintain the capacitor's voltage at a constant level. The function of MPPT is carried out in the controller of the boost chopper. More specifically, MPPT is used to change the boost chopper's duty ratio to reach MPP.

The P&O MPPT method is used as an example to explain how the duty ratio D is updated. According to (6.36), when the capacitor's voltage is maintained by the inverter's controller at a constant level, the change of the PV array's output voltage will be inverse to the change of the duty ratio D. Therefore, in Fig. 2.20, the change of voltage $U^*$ will be correspondingly replaced by the change of duty ratio. For instance, in Fig. 2.20, for the calculation path ①, we need to decrease the duty ratio D for reaching MPP. Table 6.1 summarizes the changes of duty ratio.

The change of duty ratio can also be implemented in the boost chopper's controller. An example is shown in Fig. 6.18.

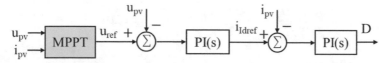

**Fig. 6.18** Example of boost chopper's controller

**Example 6.3**
An ideal PV array has 20 PV cell connected in parallel and 40 PV cell in series. The photon current under the reference condition $1000\,\text{W/m}^2$ and 298 K is 3.35 A. $\eta$ is 1.5, $k$ is $1.38047 * 10^{-23}$ J/K, $q$ is $1.60201 * 10^{-19}$ C. $C_T$ is $2.1775 * 10^{-5}$ A/K, $E_g$ is 1.14, and $I_{sref}$ is 100 pA, i.e., $100 * 10^{-12}$ A. An inverter and a chopper are used to connect the PV to an AC system, as shown in Fig. 6.17. The inverter's controller maintains the capacitor's voltage at a constant level. The initial operational point is (18.7 V, 1252 W) at $1000\,\text{W/m}^2$ and 298 K. Duty ratio of the boost chopper is 0.5 and modulation index of the inverter is 0.475.
Questions:
What is the RMS value of the AC voltage output at the initial point?
What is the duty ratio at the MPP?
What is the RMS value of the AC voltage output at the MPP when the modulation index is 0.7?

Solution: Based on the initial operational point (18.7 V, 1252 W) and duty ratio 0.5, by using (6.36), we can get the output voltage of the boost chopper, as calculated in (6.68).

$$U_{dc\_h} = \frac{1}{1-D}U_{dc\_l} = \frac{1}{1-0.5} \times 18.7 = 37.4(\text{V}). \tag{6.68}$$

Because the modulation index of the inverter is 0.475, according to (6.4), we can calculate the RMS value of the AC voltage output.

$$U_{L1} = \frac{\sqrt{3}MU_{dc}}{2\sqrt{2}} = \frac{\sqrt{3}}{2\sqrt{2}} \times 0.475 \times 37.4 = 10.879(\text{V}). \tag{6.69}$$

Based on the calculation of MPP, we can get the voltage at MPP is 32.5 V. Because the inverter's controller maintains the capacitor's voltage at 37.4 V, we can compute the value of D, which is 0.13.

Substitute the modulation index 0.7 to (6.4), we can get the RMS value of the inverter's AC voltage output.

$$U_{L1} = \frac{\sqrt{3}MU_{dc}}{2\sqrt{2}} = \frac{\sqrt{3}}{2\sqrt{2}} \times 0.7 \times 37.4 = 16.032(\text{V}). \qquad (6.70)$$

∎

## 6.5    System Integration of Energy Storage

Since the outputs of battery and supercapacitor are DC power and flywheel generates AC power, we will analyze their integration separately.

### 6.5.1    Integration of Battery and Supercapacitor

Because battery and supercapacitor produce DC power, we can directly use the single stage integration method to connect them to the AC network, as shown in Fig. 6.15, where we replace PV array with battery or supercapacitor. Compared to the single stage integration, two-stage integration enables a flexible system operation. An example of the two-stage integration method is given in Fig. 6.19, where chopper could be boost, buck, or buck-boost converter. In Fig. 6.19, the functions of PLL, outer controller, inner controller and PWM can refer to Sect. 6.2. Here, the controller of chopper is introduced for the charging mode, where battery is used as an example.

There are three typical charging methods for battery, i.e., constant current charging, constant voltage charging, and combination of constant voltage/constant current charging.

*Constant Current Charging*  In this control method, the battery is charged with a fixed current. This function can be realized through a control as given in Fig. 6.20,

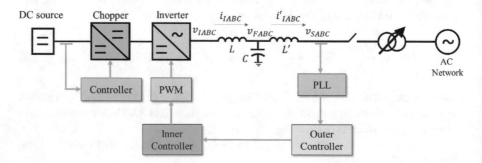

**Fig. 6.19**  Example of two-stage integration for battery and supercapacitor

**Fig. 6.20**  Constant current
charging control

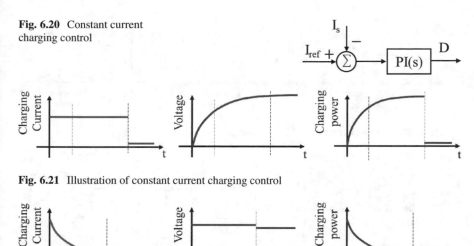

**Fig. 6.21**  Illustration of constant current charging control

**Fig. 6.22**  Illustration of constant voltage charging control

where $I_{ref}$ is a reference value of the charging current, $I_s$ is the actual charging current, and $D$ is the duty ratio of chopper.

In the constant current charging control mode, when the battery is fully charged, the charging current must be converted into a trickle to avoid damage caused by overcharging of the battery. So, the constant current charging method can fully charge the battery in a short time but must we pay attention to the charging degree of the battery. The changes of battery's current, voltage, and power under this charging control is illustrated in Fig. 6.21.

*Constant Voltage Charging*  In this control method, the battery is charged with a fixed voltage. This function can also be realized through the controller given in Fig. 6.20, where we need to replace $I_{ref}$ with the reference value of charging voltage and replace $I_s$ with the actual battery voltage.

In the constant current charging control mode, the charging current will decrease as the battery is fully charged. When the battery is fully charged, it enters the float charging mode to keep the battery fully charged. In the initial stage of charging, the initial charging current is too large due to the low voltage at the battery end.

The changes of battery's current, voltage, and power under this charging control is illustrated in Fig. 6.22.

*Combination of Constant Voltage/Constant Current Charging*  Considering the charging features of the above two methods, we can combine them together to avoid large charging current at the beginning of the constant voltage charging and avoid overcharge in the constant current charging. More specifically, we first use constant current method to charge the battery in a short time, so that we can avoid the large charging current at the beginning of the constant voltage charging. Then we shift to

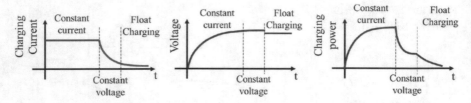

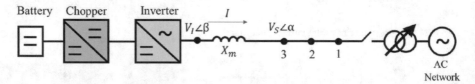

**Fig. 6.23** Illustration of combined charging control

**Fig. 6.24** System integration through inverter

constant voltage to avoid the overcharge problem in the constant current charging control method.

The changes of battery's current, voltage, and power under this charging control is illustrated in Fig. 6.23.

In the system integration of battery and supercapacitor, one important point is the power calculation. For instance, we need to figure out how much active power is charged to the storage or discharged to the AC network. To mathematically analyze the active power and reactive power, Fig. 6.24 is used. This analysis is also applicable to other inverter-integrated DERs.

In Fig. 6.24, we assume the inverter's output voltage phasor is $V_I \angle \beta$ and the integration point's voltage phasor is $V_S \angle \alpha$. The output current is $I$. The inductance between the two voltages is $X_m$.

According to KVL, we can calculate the output current, as given in (6.71).

$$I = \frac{V_I \angle \beta - V_S \angle \alpha}{jX_m}.\tag{6.71}$$

Since $V_I \angle \beta$ can be expressed as $V_I \angle \beta = V_I(\cos \beta + j \sin \beta)$ and $V_S \angle \alpha$ can be expressed as $V_S \angle \alpha = V_S(\cos \alpha + j \sin \alpha)$. Then we can rewrite (6.71) in the following equation.

$$
\begin{aligned}
I &= \frac{V_I(\cos \beta + j \sin \beta) - V_S(\cos \alpha + j \sin \alpha)}{jX_m}\\[2mm]
&= \frac{V_I \cos \beta - V_S \cos \alpha + j(V_I \sin \beta - V_S \sin \alpha)}{jX_m}\\[2mm]
&= \frac{V_I \sin \beta - V_S \sin \alpha}{jX_m} - j\frac{V_I \cos \beta - V_S \cos \alpha}{jX_m}.
\end{aligned}\tag{6.72}
$$

**Fig. 6.25** The inverter's
output voltage phasor $V_I$ is
leading the voltage phasor $V_S$

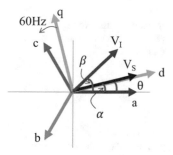

So, we can calculate the complex power delivered to the system by inverter.

$$S = V_S \cdot I^*$$

$$= V_S(\cos\alpha + j\sin\alpha) \cdot \left(\frac{V_I \sin\beta - V_S \sin\alpha}{jX_m} + j\frac{V_I \cos\beta - V_S \cos\alpha}{jX_m}\right).$$

$$(6.73)$$

Based on (6.73), the active power can be expressed as,

$$P = V_S \cos\alpha \frac{V_I \sin\beta - V_S \sin\alpha}{jX_m} - V_S \sin\alpha \frac{V_I \cos\beta - V_S \cos\alpha}{jX_m}$$

$$= \frac{1}{X_m}\left(V_S \cos\alpha V_I \sin\beta - V_S \sin\alpha V_I \cos\beta - V_S \cos\alpha V_S \sin\alpha + V_S \cos\alpha V_S \sin\alpha\right)$$

$$= \frac{1}{X_m}\left(V_S \cos\alpha V_I \sin\beta - V_S \sin\alpha V_I \cos\beta\right)$$

$$= \frac{1}{X_m}\left(V_S V_I \sin(\beta - \alpha)\right).$$

$$(6.74)$$

Based on (6.73), the reactive power can be expressed as,

$$Q = V_S \sin\alpha \frac{V_I \sin\beta - V_S \sin\alpha}{jX_m} + V_S \cos\alpha \frac{V_I \cos\beta - V_S \cos\alpha}{jX_m}$$

$$= \frac{1}{X_m}\left(V_S \sin\alpha V_I \sin\beta + V_S \cos\alpha V_I \cos\beta - V_S \sin\alpha V_S \sin\alpha - V_S \cos\alpha V_S \cos\alpha\right)$$

$$= \frac{1}{X_m}\left(V_S V_I \cos(\beta - \alpha) - V_S^2\right).$$

$$(6.75)$$

*Active Power Analysis* From (6.74), we can see the active power output or input depends on the value of $\beta - \alpha$, i.e., the phasor angle difference. Depending on the value of $\beta - \alpha$, theoretically, there are three cases.

1. $\beta - \alpha > 0$. In this case, the inverter's output voltage phasor $V_I$ is leading the voltage phasor $V_S$, as illustrated in Fig. 6.25.

**Fig. 6.26** Scenario when inverter's output voltage phasor $V_I$ leads the voltage phasor $V_S$

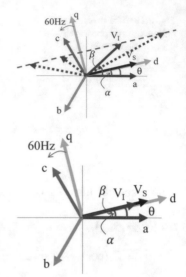

**Fig. 6.27** The inverter's output voltage phasor $V_I$ is in phase with the voltage phasor $V_S$

From Fig. 6.25, we can see in this case, the active power output is positive, namely, the energy storage unit is discharging power to the AC system.

$$P = \frac{1}{X_m}\left(V_S V_I \sin(\beta - \alpha)\right) > 0. \tag{6.76}$$

From Fig. 6.25, we can also see as long as the voltage phasor $V_I$ is along the dash line, as illustrated in Fig. 6.26, $V_I \sin(\beta - \alpha)$ will be the same, that is the inverter will generate the same amount of active power.

2. $\beta - \alpha = 0$. In this case, the inverter's output voltage phasor $V_I$ is in phase with the voltage phasor $V_S$, as illustrated in Fig. 6.27.

From Fig. 6.27, we can see in this case, the active power output is zero, namely, the energy storage unit neither charges or discharges power.

$$P = \frac{1}{X_m}\left(V_S V_I \sin(\beta - \alpha)\right) = 0. \tag{6.77}$$

From Fig. 6.27, we can also see as long as the voltage phasor $V_I$ is in phase with $V_S$, that is along the dash line, as illustrated in Fig. 6.28, $V_I \sin(\beta - \alpha)$ will be zero. In other word, along the dash line in Fig. 6.28, the inverter will not generate or charge active power.

3. $\beta - \alpha < 0$. In this case, the inverter's output voltage phasor $V_I$ is lagging the voltage phasor $V_S$, as illustrated in Fig. 6.29.

From Fig. 6.29, we can see in this case, the active power output is negative, which means the energy storage unit is in the charging mode.

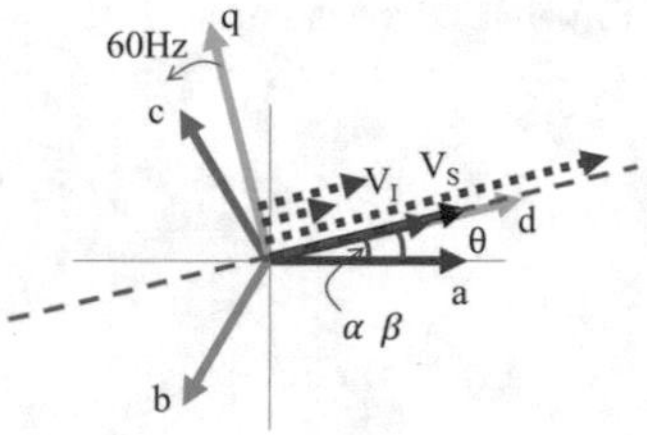

**Fig. 6.28** Scenarios when the voltage phasor $V_I$ is in phase with $V_S$

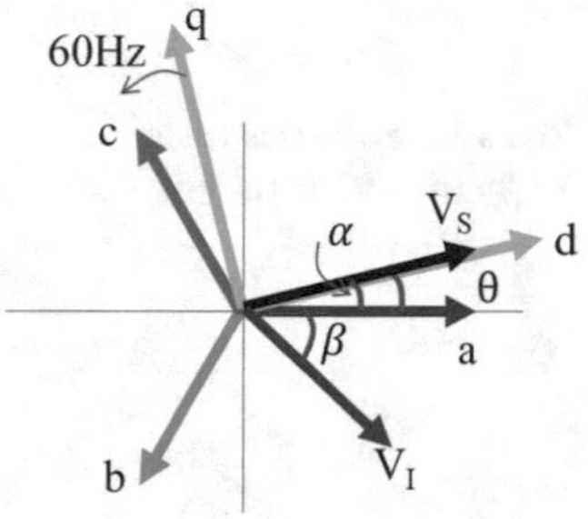

**Fig. 6.29** The inverter's output voltage phasor $V_I$ is lagging the voltage phasor $V_S$

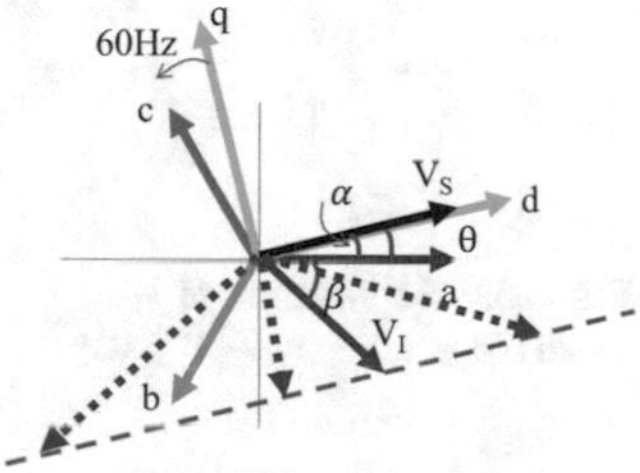

**Fig. 6.30** Scenarios when the voltage phasor $V_I$ lags the voltage phasor $V_S$

$$P = \frac{1}{X_m}\left(V_S V_I \sin(\beta - \alpha)\right) < 0. \tag{6.78}$$

From Fig. 6.29, we can also see as long as the voltage phasor $V_I$ is along the dash line, as illustrated in Fig. 6.30, $V_I \sin(\beta - \alpha)$ will have the same amount. In other word, along the dash line in Fig. 6.30, the inverter will charge the same amount of active power.

*Reactive Power Analysis* From (6.75), we can see the reactive power output or input depends on the value of $V_I \cos(\beta - \alpha) - V_S$. Depending on the value of $V_I \cos(\beta - \alpha) - V_S$, theoretically, there are also three cases.

1. $V_I \cos(\beta - \alpha) - V_S > 0$. In this case, the projection of the inverter's output voltage phasor $V_I$ to the d-axis (in phase with $V_S$) is larger than the voltage phasor $V_S$, as illustrated in Fig. 6.31.

    From Fig. 6.31, we can see in this case, the reactive power output is positive, namely, the inverter generates reactive power to the AC system.

**Fig. 6.31** Explanation of
$V_I \cos{(\beta - \alpha)} - V_S > 0$

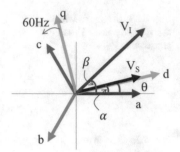

**Fig. 6.32** Scenarios when
$V_I \cos{(\beta - \alpha)} - V_S > 0$

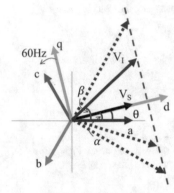

**Fig. 6.33** Explanation of
$V_I \cos{(\beta - \alpha)} - V_S = 0$

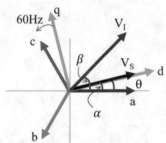

$$Q = \frac{1}{X_m}\left(V_S V_I \cos{(\beta - \alpha)} - V_S^2\right) > 0. \tag{6.79}$$

From Fig. 6.31, we can also see as long as the voltage phasor $V_I$ is along the dash line, as illustrated in Fig. 6.32, $V_I \cos{(\beta - \alpha)} - V_S$ will have the same value. In other word, along the dash line in Fig. 6.32, the inverter will generate the same amount of reactive power.

2. $V_I \cos{(\beta - \alpha)} - V_S = 0$. In this case, the projection of the inverter's output voltage phasor $V_I$ to the d-axis is equal to the voltage phasor $V_S$, as illustrated in Fig. 6.33.

From Fig. 6.33, we can see in this case, the reactive power output is zero, namely, the inverter does not exchange reactive power with the AC system.

**Fig. 6.34** Scenarios when
$V_I \cos{(\beta - \alpha)} - V_S = 0$

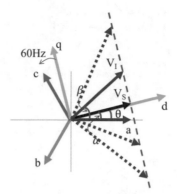

**Fig. 6.35** Explanation of
$V_I \cos{(\beta - \alpha)} - V_S < 0$

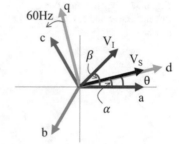

$$Q = \frac{1}{X_m}\left(V_S V_I \cos{(\beta - \alpha)} - V_S^2\right) = 0. \tag{6.80}$$

From Fig. 6.33, we can also see as long as the voltage phasor $V_I$ is along the dash line, as illustrated in Fig. 6.34, $V_I \cos{(\beta - \alpha)} - V_S$ will be zero. In other word, along the dash line in Fig. 6.34, the inverter will generate no reactive power.

3. $V_I \cos{(\beta - \alpha)} - V_S < 0$. In this case, the projection of the inverter's output voltage phasor $V_I$ to the d-axis is smaller to the voltage phasor $V_S$, as illustrated in Fig. 6.35.

From Fig. 6.35, we can see in this case, the reactive power output is negative, namely, the inverter absorbs reactive power from the AC system.

$$Q = \frac{1}{X_m}\left(V_S V_I \cos{(\beta - \alpha)} - V_S^2\right) < 0. \tag{6.81}$$

From Fig. 6.35, we can also see as long as the voltage phasor $V_I$ is along the dash line, as illustrated in Fig. 6.36, $V_I \cos{(\beta - \alpha)} - V_S$ will have the same value. In other word, along the dash line in Fig. 6.36, the inverter will generate the same amount of reactive power.

From the above analysis, we can see through controlling the output voltage of the inverter, the active power and reactive power can be adjusted. There are several applications, e.g., adjust system voltage through controlling reactive power,

**Fig. 6.36** Scenarios when
$V_I \cos(\beta - \alpha) - V_S < 0$

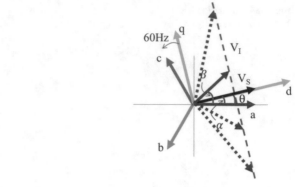

**Fig. 6.37** Explanation of
inverter active and reactive
power

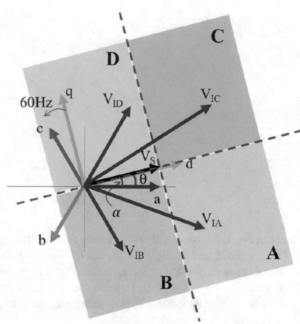

charge/discharge energy storage units, smooth the output of DERs, etc. In summary,
the plane can be divided into four regions, as illustrated in Fig. 6.37.

*Region A* In this region, the inverter's output voltage phasor, e.g., $V_I A$, is lagging
$V_S$, which means the inverter receives active power. Since $V_I \cos(\beta - \alpha) - V_S$ is
larger than zero, the inverter generates reactive power to the AC system.

*Region B* In this region, the inverter's output voltage phasor, e.g., $V_I A$, is lagging
$V_S$, which means the inverter receives active power. Since $V_I \cos(\beta - \alpha) - V_S$ is
smaller than zero, the inverter absorbs reactive power from the AC system.

*Region C* In this region, the inverter's output voltage phasor, e.g., $V_I A$, is leading
$V_S$, which means the inverter generates active power to the AC system. Since

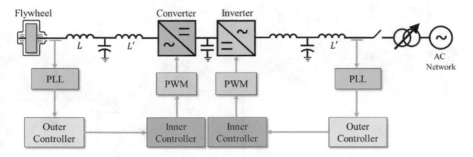

**Fig. 6.38** Integration of flywheel system into the AC network

$V_I \cos(\beta - \alpha) - V_S$ is larger than zero, the inverter generates reactive power to the AC system.

*Region D*  In this region, the inverter's output voltage phasor, e.g., $V_I A$, is leading $V_S$, which means the inverter generates active power to the AC system. Since $V_I \cos(\beta - \alpha) - V_S$ is smaller than zero, the inverter absorbs reactive power from the AC system.

## 6.5.2   Integration of Flywheel

Because the flywheel system produces AC power which is usually not nominal frequency, two-stage integration method is recommended to connect the flywheel system to the AC network, as shown in Fig. 6.38, which can enables a flexible system operation. In Fig. 6.38, the basic functions of PLL, outer controller, inner controller and PWM can refer to Sect. 6.2.

Here, the charging method for the flywheel side converter is discussed. As introduced in Sect. 3.2, constant torque method and constant power are two basic charging methods.

*Constant Power Control*  Based on (3.8), we can get the flywheel's power computation, as shown in (6.82). It also verifies the calculation derived in (3.13).

$$P = \frac{dE}{dt} = \frac{d}{dt}\left(\frac{1}{2}J_F\omega_g^2\right) = J_F\omega_g\frac{d\omega_g}{dt}. \tag{6.82}$$

Applying Laplace transformation to (6.82), we can get $P = J_F\omega_g s\omega_g$. Therefore, in the constant power control, we can set the reference value for the charging power, that is,

$$P_{ref} = J_F\omega_{gref}s\omega_{gref}. \tag{6.83}$$

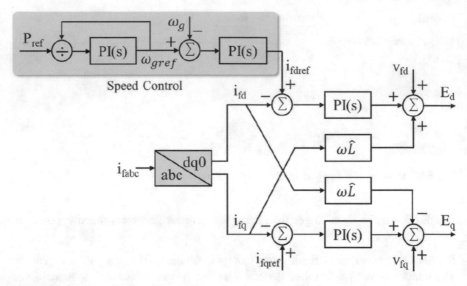

**Fig. 6.39** Controller for the flywheel side converter

Then we can rewrite (6.83) to get the reference value of angular velocity, as below.

$$\omega_{gref} = \frac{P_{ref}}{\omega_{gref}} \cdot \frac{1}{J_{FS}}. \tag{6.84}$$

Correspondingly, the controller for the flywheel side converter can be revised as shown in Fig. 6.39.

*Constant Torque Control* Since $P$ can also be calculated through torque, i.e., $P = T\omega_g$, we then have $T = J_{FS}\omega_g$. Therefore, in the constant torque control, we can set the reference value for the charging torque as shown in (6.85). Correspondingly, the controller for the flywheel side converter should be revised according to (6.85).

$$T_{ref} = J_{FS}\omega_{gref}. \tag{6.85}$$

## 6.6   System Integration of Micro-Turbine

As introduced, microturbines rotate at over 30,000 Revolutions Per Minute (RPM). The high speed generates very high frequency power. Therefore, two-stage integration method is also recommended to connect the microturbine system to the AC network, as shown in Fig. 6.40, which can enables a flexible system operation. In Fig. 6.40, the basic functions of PLL, outer controller, inner controller and PWM can refer to Sect. 6.2. Note that the function of inverter control is usually to keep

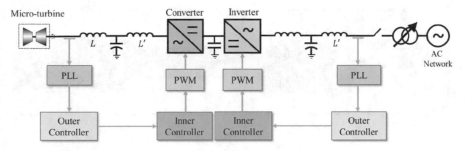

**Fig. 6.40**  Integration of microturbine system into the AC network

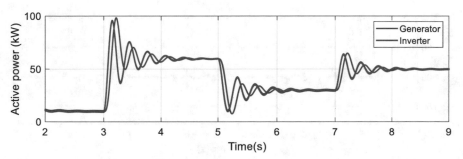

**Fig. 6.41**  Active power outputs of generator and inverter

the voltage of DC capacitor at a constant level. One simulation example given in Example 6.4 is used to demonstrate the control and operations of the microturbine system.

---

**Example 6.4**

A microturbine is connected to the AC network through back-to-back power electronics interface given in Fig. 6.40. The DC capacitor's voltage is 800 V. At 3.0 s, the microturbine's active power output is dispatched from 10 to 60 kW, then to 30 kW at 5.0 s, and then to 50 kW at 7.0 s.

Questions:

Please demonstrate the changes of the active power of generator and inverter, inverter's output AC current, capacitor's DC voltage, and the AC system's frequency.

---

Simulation results are shown in Figs. 6.41, 6.42, 6.43, and 6.44.

From the active power shown in Fig. 6.41 and current shown in Fig. 6.42, we can see that,

- The microturbine can generate the required amount of active power.

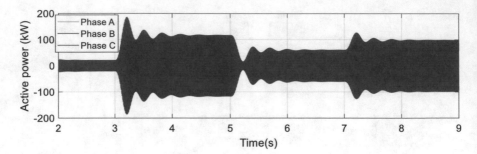

**Fig. 6.42** Three-phase current outputs of inverter

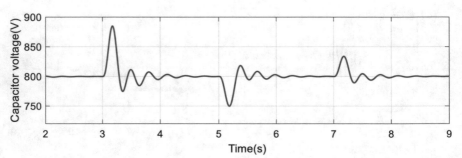

**Fig. 6.43** Capacitor's voltage

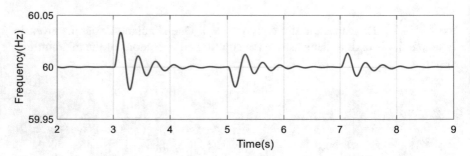

**Fig. 6.44** System's frequency

- There is a dynamic differences between the generator's and inverter's active power output. And the change of generator's active power is sort of leading that of inverter's active power. This difference is caused by the DC capacitor, which helps smooth the dynamics of microturbine's dynamic power output.
- In the steady state, the generator's active power output is equal to the inverter's output, which means there is no charging or discharging in the DC capacitor.
- The inverter's AC current can correspondingly change to produce the required amount of active power.

From the capacitor's voltage shown in Fig. 6.43 and system's frequency shown in Fig. 6.44, we can see that,

- The capacitor's voltage can maintain at the given value, in this case, it is 800 V.
- The changes of active power output from microturbine cause a frequency dynamics, and eventually the system's frequency can maintain at 60 Hz. ∎

## 6.7 System Integration of Wind Generation

Since DFIG is a popular wind turbine, here we mainly introduce the system integration of DFIG. The system integration of DFIG is shown in Fig. 5.12. A control system is employed to control the machine side converter and grid side converter for regulating the active and reactive power (by regulating the current flowing in the rotor winding) to extract the maximum possible power from the wind and to regulate the reactive power output of the generator. There are two typical control methods [27], namely vector control or field-oriented control, and direct torque control (DTC), where vector control is the predominant control.

Vector control allows decoupling of active and reactive power control [28], i.e., active power can be independently controlled without affecting reactive power output and vice versa. There is a wide generator shaft speed range of up to 30% above and below rated speed.

To apply the vector control method to control real and reactive power output, it is necessary to understand the behavior of the wound rotor induction machine because it is frequently used in DFIG.

### 6.7.1 Equivalent Circuit of Wound Rotor Induction Generator

In the wound rotor induction generator, the stator's and rotor's three-phase magnetic fluxes are illustrated in Fig. 6.45. Based on Fig. 6.45, we can correspondingly develop the stator's and rotor's electric circuits, as given in Figs. 6.46 and 6.47, respectively.

Based on Fig. 6.46, applying KVL to each phase in the stator, we can have the following equation,

$$\begin{cases} v_{As}(t) = r_s i_{As}(t) + e_{As} = r_s i_{As}(t) + \dfrac{d\Phi_{As}}{dt} \\[2mm] v_{Bs}(t) = r_s i_{Bs}(t) + e_{Bs} = r_s i_{Bs}(t) + \dfrac{d\Phi_{Bs}}{dt} \\[2mm] v_{Cs}(t) = r_s i_{Cs}(t) + e_{Cs} = r_s i_{Cs}(t) + \dfrac{d\Phi_{Cs}}{dt}, \end{cases} \qquad (6.86)$$

**Fig. 6.45** Stator's and rotor's three-phase magnetic fluxes

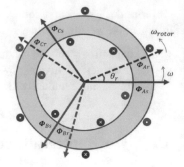

**Fig. 6.46** Stator's electric circuits

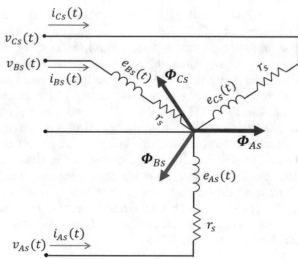

where,

- $\Phi_{As}$, $\Phi_{Bs}$, and $\Phi_{Cs}$ are the linkage fluxes of stator's phase A, phase B, and phase C, respectively.
- $i_{As}(t)$, $i_{Bs}(t)$, and $i_{Cs}(t)$ are the stator's magnetizing currents.

Similarly, based on Fig. 6.47, applying KVL to each phase in the rotor, we can have the following equation,

$$
\begin{cases}
v_{Ar}(t) = r_r i_{Ar}(t) + e_{Ar} = r_r i_{Ar}(t) + \dfrac{d\Phi_{Ar}}{dt} \\[2mm]
v_{Br}(t) = r_r i_{Br}(t) + e_{Br} = r_r i_{Br}(t) + \dfrac{d\Phi_{Br}}{dt} \\[2mm]
v_{Cr}(t) = r_r i_{Cr}(t) + e_{Cr} = r_r i_{Cr}(t) + \dfrac{d\Phi_{Cr}}{dt},
\end{cases}
\tag{6.87}
$$

where,

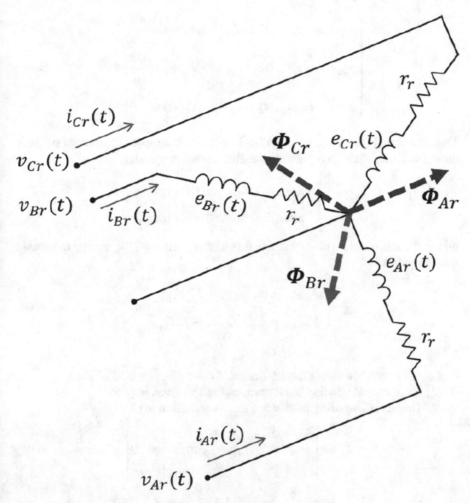

**Fig. 6.47** Rotor's electric circuits

- $\Phi_{Ar}$, $\Phi_{Br}$, and $\Phi_{Cr}$ are the linkage fluxes of rotor's phase A, phase B, and phase C, respectively.
- $i_{Ar}(t)$, $i_{Br}(t)$, and $i_{Cr}(t)$ are the rotor's magnetizing currents.

To simplify the analysis, we convert the rotor side's calculation to the stator side. Hence, (6.87) can be updated to,

$$\begin{cases} v'_{Ar}(t) = r'_r i'_{Ar}(t) + e'_{Ar} = r'_r i'_{Ar}(t) + \dfrac{d\Phi'_{Ar}}{dt} \\[2mm] v'_{Br}(t) = r'_r i'_{Br}(t) + e'_{Br} = r'_r i'_{Br}(t) + \dfrac{d\Phi'_{Br}}{dt} \\[2mm] v'_{Cr}(t) = r'_r i'_{Cr}(t) + e'_{Cr} = r'_r i'_{Cr}(t) + \dfrac{d\Phi'_{Cr}}{dt}. \end{cases} \tag{6.88}$$

Then (6.86) and (6.88) can be rewritten in the following matrix format to simplify the expressions, where $p$ represents the differentiation operator.

$$\begin{cases} v_{ABCs} = r_s i_{ABCs} + p\Phi_{ABCs} \\[1mm] v'_{ABCr} = r'_r i'_{ABCr} + p\Phi'_{ABCr} \end{cases} \tag{6.89}$$

where, the stator's and rotor's linkage fluxes $\Phi_{ABCs}$ and $\Phi'_{ABCr}$ can be calculated through (6.90).

$$\begin{cases} \Phi_{ABCs} = L_s i_{ABCs} + L'_{sr} i'_{ABCr} \\[1mm] \Phi'_{ABCr} = L'^T_{sr} i'_{ABCs} + L'_r i'_{ABCr} \end{cases} \tag{6.90}$$

where,

- $L_s$ is the stator's winding inductance matrix and given in (6.91).
- $L'_{sr}$ is the mutual winding inductance matrix and given in (6.92).
- $L'_r$ is the rotor's winding inductance matrix and given in (6.93).

$$L_s = \begin{bmatrix} L_{ls} + L_{ms} & -\frac{L_{ms}}{2} & -\frac{L_{ms}}{2} \\[2mm] -\frac{L_{ms}}{2} & L_{ls} + L_{ms} & -\frac{L_{ms}}{2} \\[2mm] -\frac{L_{ms}}{2} & \frac{L_{ms}}{2} & L_{ls} + L_{ms} \end{bmatrix} \tag{6.91}$$

$$\cdot\; L'_{sr} = L_{ms} \begin{bmatrix} cos(\theta_r) & cos(\theta_r + 120°) & cos(\theta_r - 120°) \\[2mm] cos(\theta_r - 120°) & cos(\theta_r) & cos(\theta_r + 120°) \\[2mm] cos(\theta_r + 120°) & cos(\theta_r - 120°) & cos(\theta_r) \end{bmatrix} \tag{6.92}$$

$$L'_r = \begin{bmatrix} L'_{lr} + L_{ms} & -\frac{L_{ms}}{2} & -\frac{L_{ms}}{2} \\[2mm] -\frac{L_{ms}}{2} & L'_{lr} + L_{ms} & -\frac{L_{ms}}{2} \\[2mm] -\frac{L_{ms}}{2} & \frac{L_{ms}}{2} & L'_{lr} + L_{ms} \end{bmatrix} \tag{6.93}$$

Substitute (6.90) to (6.89), we can get,

$$\begin{cases} v_{ABCs} = (r_s + pL_s) i_{ABCs} + pL'_{sr} i'_{ABCr} \\[1mm] v'_{ABCr} = pL'^T_{sr} i_{ABCs} + (r'_r + pL'_r) i'_{ABCr}. \end{cases} \tag{6.94}$$

## 6.7.2 Equivalent Model in the dq0 Frame

In the stationary $abc$ reference frame, (6.94) shows that machine parameters such as inductance are time-varying. The equivalent circuit in the stationary $abc$ reference frame can then be transformed to the equivalent in the rotating $dq0$ reference frame, to make machine parameters, such as inductance, time-invariant. Let's take $v_{ABCs} = (r_s + pL_s)i_{ABCs} + pL'_{sr}i'_{ABCr}$ as an example to do the $dq0$ transformation. Constant amplitude $dq0$ transformation, as shown in (6.16), is adopted. Therefore, we can have the following expressions of $v_{ABCs}$ and $v_{dq0s}$.

$$\begin{cases} v_{ABCs} = P_\theta^{-1} v_{dq0s} \\ v_{dq0s} = P_\theta v_{ABCs}. \end{cases} \tag{6.95}$$

So that the equation $v_{ABCs} = (r_s + pL_s)i_{ABCs} + pL'_{sr}i'_{ABCr}$ can be rewritten as,

$$P_\theta^{-1} v_{dq0s} = (r_s + pL_s)P_\theta^{-1} i_{dq0s} + pL'_{sr}P_\theta^{-1} i'_{dq0r}. \tag{6.96}$$

Multiplying $P_\theta$ in (6.96), the following equation can be obtained.

$$\begin{aligned} v_{dq0s} &= P_\theta(r_s + pL_s)P_\theta^{-1} i_{dq0s} + P_\theta p L'_{sr} P_\theta^{-1} i'_{dq0r} \\ &= P_\theta r_s P_\theta^{-1} i_{dq0s} + P_\theta L_s p P_\theta^{-1} i_{dq0s} + P_\theta L'_{sr} p P_\theta^{-1} i'_{dq0r} \\ &= A + P_\theta L_s B + P_\theta L'_{sr} C. \end{aligned} \tag{6.97}$$

where,

- $A = P_\theta r_s P_\theta^{-1} i_{dq0s}.$
- $B = p P_\theta^{-1} i_{dq0s}.$
- $C = p P_\theta^{-1} i'_{dq0r}.$

*Computation of A* In (6.97), $A$ can be computed below.

$$A = P_\theta r_s P_\theta^{-1} i_{dq0s} = r_s i_{dq0s} \tag{6.98}$$

*Computation of B* In (6.97), $B$ can be computed below.

$$B = p P_\theta^{-1} i_{dq0s} = (p P_\theta^{-1})i_{dq0s} + P_\theta^{-1}(p i_{dq0s}) \tag{6.99}$$

where,

$$pP_\theta^{-1} = \frac{d}{dt} \begin{bmatrix} \cos\theta & -\sin\theta & 1 \\ \cos(\theta - \frac{2\pi}{3}) & -\sin(\theta - \frac{2\pi}{3}) & 1 \\ \cos(\theta + \frac{2\pi}{3}) & -\sin(\theta + \frac{2\pi}{3}) & 1 \end{bmatrix}$$

$$= \begin{bmatrix} -\omega\sin\theta & -\omega\cos\theta & 0 \\ -\omega\sin(\theta - \frac{2\pi}{3}) & -\omega\cos(\theta - \frac{2\pi}{3}) & 0 \\ -\omega\sin(\theta + \frac{2\pi}{3}) & -\omega\cos(\theta + \frac{2\pi}{3}) & 0 \end{bmatrix}$$

$$= -\begin{bmatrix} \cos\theta & -\sin\theta & 1 \\ \cos(\theta - \frac{2\pi}{3}) & -\sin(\theta - \frac{2\pi}{3}) & 1 \\ \cos(\theta + \frac{2\pi}{3}) & -\sin(\theta + \frac{2\pi}{3}) & 1 \end{bmatrix} \begin{bmatrix} 0 & \omega & 0 \\ -\omega & 0 & 0 \\ 0 & 0 & 0 \end{bmatrix} \qquad (6.100)$$

$$= -P_\theta^{-1} \begin{bmatrix} 0 & \omega & 0 \\ -\omega & 0 & 0 \\ 0 & 0 & 0 \end{bmatrix}$$

Therefore, $P_\theta L_s B$ is calculated below.

$$P_\theta L_s B = -P_\theta L_s P_\theta^{-1} \begin{bmatrix} 0 & \omega & 0 \\ -\omega & 0 & 0 \\ 0 & 0 & 0 \end{bmatrix} i_{dq0s} + P_\theta L_s P_\theta^{-1}(p i_{dq0s})$$

$$= -L_s \begin{bmatrix} 0 & \omega & 0 \\ -\omega & 0 & 0 \\ 0 & 0 & 0 \end{bmatrix} i_{dq0s} + L_s(p i_{dq0s})$$

$$= -L_s \begin{bmatrix} -\omega i_{qs} \\ \omega i_{ds} \\ 0 \end{bmatrix} + L_s(p i_{dq0s}) \qquad (6.101)$$

$$= L_s \begin{bmatrix} -\omega i_{qs} + \frac{di_{ds}}{dt} \\ \omega i_{ds} + \frac{di_{qs}}{dt} \\ 0 + \frac{di_{0s}}{dt} \end{bmatrix}$$

*Computation of C* Similar to the computation of $P_\theta L_s B$, in (6.97), $P_\theta L'_{sr} C$ can be computed below.

$$P_\theta L'_{sr} C = L'_{sr} \begin{bmatrix} -\omega i'_{qr} + \frac{di'_{dr}}{dt} \\ \omega i'_{dr} + \frac{di'_{qr}}{dt} \\ 0 + \frac{di'_{0r}}{dt} \end{bmatrix} \qquad (6.102)$$

Then based on (6.98), (6.101), and (6.102), we can get,

$$\begin{bmatrix} v_{ds} \\ v_{qs} \\ v_{0s} \end{bmatrix} = \begin{bmatrix} r_s i_{ds} \\ r_s i_{qs} \\ r_s i_{0s} \end{bmatrix} + L_s \begin{bmatrix} -\omega i_{qs} + \frac{di_{ds}}{dt} \\ \omega i_{ds} + \frac{di_{qs}}{dt} \\ 0 + \frac{di_{0s}}{dt} \end{bmatrix} + L'_{sr} \begin{bmatrix} -\omega i'_{qr} + \frac{di'_{dr}}{dt} \\ \omega i'_{dr} + \frac{di'_{qr}}{dt} \\ 0 + \frac{di'_{0r}}{dt} \end{bmatrix} \qquad (6.103)$$

Therefore, we can get the following stator voltage in the $dq$ frame,

$$\begin{cases} v_{ds} = r_s i_{ds} - \omega \Phi_{qs} + \dfrac{d\Phi_{ds}}{dt} \\[3mm] v_{qs} = r_s i_{qs} + \omega \Phi_{ds} + \dfrac{d\Phi_{qs}}{dt}. \end{cases} \qquad (6.104)$$

Similarly, we can get the $dq$ transformations for the rotor voltage shown in (6.105), stator's flux linkage in (6.106), and rotor's flux linkage in (6.107), respectively.

$$\begin{cases} v'_{dr} = r'_r i'_{dr} - (\omega - \omega_r)\Phi'_{qr} + \dfrac{d\Phi'_{dr}}{dt} \\[3mm] v'_{qr} = r'_r i'_{qr} + (\omega - \omega_r)\Phi'_{dr} + \dfrac{d\Phi'_{qr}}{dt}. \end{cases} \qquad (6.105)$$

$$\begin{cases} \Phi_{ds} = (L_{ls} + L_M)i_{ds} + L_M i'_{dr} \\[2mm] \Phi_{qs} = (L_{ls} + L_M)i_{qs} + L_M i'_{qr}. \end{cases} \qquad (6.106)$$

$$\begin{cases} \Phi'_{dr} = L_M i_{ds} + (L'_{lr} + L_M)i'_{dr} \\[2mm] \Phi'_{qr} = L_M i_{qs} + (L'_{lr} + L_M)i'_{qr}. \end{cases} \qquad (6.107)$$

Substitute (6.106) into (6.104), we can have the following stator voltage,

$$\begin{cases} v_{ds} = r_s i_{ds} - \omega \Phi_{qs} + \dfrac{d}{dt}\big((L_{ls} + L_M)i_{ds} + L_M i'_{dr}\big) \\[3mm] v_{qs} = r_s i_{qs} + \omega \Phi_{ds} + \dfrac{d}{dt}\big((L_{ls} + L_M)i_{qs} + L_M i'_{qr}\big). \end{cases} \qquad (6.108)$$

Substitute (6.107) into (6.105), we can have the following rotor voltage,

$$\begin{cases} v'_{dr} = r'_r i'_{dr} - (\omega - \omega_r)\Phi'_{qr} + \dfrac{d}{dt}\big(L_M i_{ds} + (L'_{lr} + L_M)i'_{dr}\big) \\[3mm] v'_{qr} = r'_r i'_{qr} + (\omega - \omega_r)\Phi'_{dr} + \dfrac{d}{dt}\big(L_M i_{qs} + (L'_{lr} + L_M)i'_{qr}\big). \end{cases} \qquad (6.109)$$

Based on (6.108) and (6.110), Figs. 6.48 and 6.49 can be used to represent the two equations.

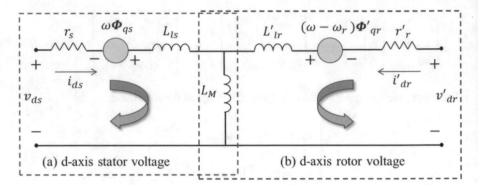

**Fig. 6.48**  Equivalent circuit to represent $d$-axis calculation

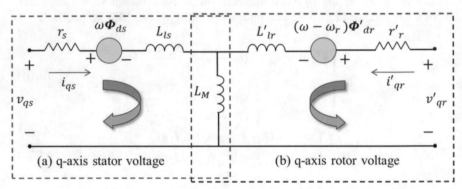

**Fig. 6.49**  Equivalent circuit to represent $q$-axis calculation

**Fig. 6.50**  Total magnetic
flux phasor representation

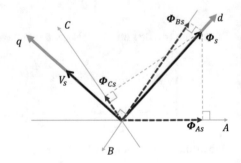

### 6.7.3  Vector Control

In the $dq0$ transformation as given in (6.17), the angle $\theta$ is playing an important role. The currents flowing in the stator are assumed to be balanced. Considering it, these currents produce a resultant stator magnetic field which has a constant magnitude and is rotating at synchronous speed.

The total magnetic flux phasor $\Phi_s$ is given in Fig. 6.50. Its projections to A-axis, B-axis, and C-axis are $\Phi_{As}$, $\Phi_{Bs}$, and $\Phi_{Cs}$, respectively. Considering the voltage phasor $V_s$ is leading the magnetic flux phasor $\Phi_s$ by 90°, we can get the direction of the voltage phasor $V_s$. Considering (6.104), to simplify the control system, we can have the $d$-axis in phase with the magnetic flux phasor $\Phi_s$ and $q$-axis is leading $d$-axis by 90°. In this condition, the $d$-axis and $q$-axis magnetic fluxes derived in (6.106) can be simplified as (6.110).

$$
\begin{cases}
\Phi_{ds} = \Phi_s \\
\Phi_{qs} = 0.
\end{cases}
\tag{6.110}
$$

When the magnetic flux phasor $\Phi_s$ is in phase with the d-axis, its value will not change over time, i.e., $\frac{d\Phi_{ds}}{dt} = 0$. Therefore, (6.104) can be correspondingly simplified as (6.111), where $r_s$ is ignored.

$$
\begin{cases}
v_{ds} = r_s i_{ds} - \omega\Phi_{qs} + \dfrac{d\Phi_{ds}}{dt} = 0 \\[2mm]
v_{qs} = r_s i_{qs} + \omega\Phi_{ds} + \dfrac{d\Phi_{qs}}{dt} = \omega\Phi_{ds} = \omega\Phi_s.
\end{cases}
\tag{6.111}
$$

From (6.111), we can see there only exists $q$-axis projection of the voltage phasor, which can verify the setting of $dq$-axis in Fig. 6.50. From (6.110), we can also simply the dq current $i_{ds}$ and $i_{qs}$.

$$
\begin{cases}
i_{ds} = \dfrac{\Phi_s - L_M i'_{dr}}{L_{ls} + L_M} \\[4mm]
i_{qs} = -\dfrac{L_M}{L_{ls} + L_M} i'_{qr}.
\end{cases}
\tag{6.112}
$$

Considering (6.111), (6.112) and the $dq0$ frame power calculation given in (6.22), for the balanced system, 0-axis value will be zero, then the active power and reactive power can be simplified as,

$$
\begin{cases}
P_s = \dfrac{3}{2}(v_{ds} i_{ds} + v_{qs} i_{qs}) = \dfrac{3}{2} v_{qs} i_{qs} = -\dfrac{3\omega\Phi_s}{2}\left(\dfrac{L_M}{L_{ls} + L_M} i'_{qr}\right) \\[4mm]
Q_s = \dfrac{3}{2}(v_{ds} i_{qs} - v_{qs} i_{ds}) = -\dfrac{3}{2} v_{qs} i_{ds} = -\dfrac{3\omega\Phi_s}{2}\left(\dfrac{\Phi_s - L_M i'_{dr}}{L_{ls} + L_M}\right).
\end{cases}
\tag{6.113}
$$

From (6.113), we can see through controlling the rotor's current $i'_{qr}$, we can adjust the stator's active power out $P_s$. Through controlling the rotor's current $i'_{dr}$, we can adjust the stator's reactive power out $Q_s$. Therefore, vector control allows decoupling of active and reactive power control. In practice, a control system is

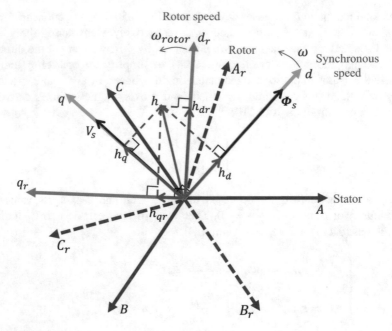

**Fig. 6.51** Frames used for vector control

employed to regulate the stator's active and reactive power outputs by regulating the current flowing in the rotor winding.

Several frames have been discussed. Figure 6.51 summarizes their relations.

- In the stator side, its $ABC$ frame is stationary.
- In the stator side, its $dq$ frame is rotating at the synchronous speed $\omega$. The stator's total magnetic flux $\Phi_s$ is in phase with $d$-axis. So that the voltage phasor is in phase with $q$-axis.
- In the rotor side, its $ABC_r$ frame is stationary.
- In the rotor side, its $dq_r$ frame is rotating at the rotor speed $\omega_{rotor}$.
- Assume the control signal is $h$ as illustrated in Fig. 6.51. Through dq transformation, we can get $h_d$ and $h_q$. Note that $h_d$ and $h_q$ are still in the stator's $dq$ frame, which is rotating at the synchronous speed $\omega$. We then need to convtert them to the rotor's $dq_r$ frame to get $h_{dr}$ and $h_{qr}$. $h_{dr}$ and $h_{qr}$ will be used in the rotor side controller.

## Problems

**6.1** Describe the function of three-phase inverter.

**6.2** Describe the function of boost chopper, buck chopper, and buck-boost chopper.

**6.3** Describe the vector control in the system integration of wind turbine.

**6.4** In the wind generation system, prove the rotor side voltage can be expressed in (6.105).

**6.5** In the wind generation system, prove the stator's flux linkage can be expressed in (6.106) and rotor's flux linkage can be expressed in (6.107).

**6.6** A buckboost chopper as shown in Fig. 6.14 is used to connect a DC voltage source to a power load. The DC voltage is 20 V and the current is 50 A.

Question 1: When the duty ratio is equal to 0.2, what is the output voltage of the buckboost chopper? What is the power load's DC current? What is the function of the buckboost chopper?

Question 2: When the duty ratio is equal to 0.4, what is the output voltage of the buckboost chopper? What is the power load's DC current? What is the function of the buckboost chopper?

Question 3: When the duty ratio is equal to 0.5, what is the output voltage of the buckboost chopper? What is the power load's DC current? What is the function of the buckboost chopper?

Question 4: When the duty ratio is equal to 0.8, what is the output voltage of the buckboost chopper? What is the power load's DC current? What is the function of the buckboost chopper?

**6.7** In Fig. 6.24, a battery is connected to the AC network through a buck-boost converter and an inverter. It delivers 100 kW active power and 40 kVAR reactive power to the main grid. The voltage at bus 3 is $0.4\angle15°$ kV. $X_m$ is 85 mH. The modulation index of inverter is 0.8. The duty ratio of converter is 0.4.

Question 1: What is the output voltage of battery in the above scenario?

Question 2: If we need to charge the battery, should $q$-axis modulation index be positive or negative?

Question 3: If we need to discharge the battery, should $q$-axis modulation index be positive or negative?

**6.8** In Fig. 6.24, a battery is connected to the AC network through a buck-boost converter and an inverter. The voltage at bus 3 is $0.3\angle30°$ kV. $X_m$ is 60 mH. The modulation index of inverter is 0.9. The duty ratio of converter is 0.6. The output voltage of battery is 600 V. It delivers 100 kW active power and 40 kVAR reactive power to the main grid.

Question 1: What is the power output of the battery?

Question 2: Does the battery charge or discharge power? Why?

Question 3: Does the buck-boost converter work in the buck mode or boost mode? Why?

**6.9** In Fig. 6.24, a battery is connected to the AC network through a buck-boost converter and an inverter. The inverter's output voltage phasor is $V_I$, as shown in Fig. 6.52.

Question 1: What is the operational mode of the battery, charging or discharging power?

**Fig. 6.52** Inverter's output
voltage for Problem 6.9

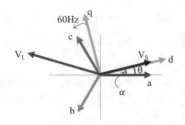

Question 2: Does the inverter deliver or absorb reactive power?

Question 3: In order to charge the battery, what conditions should we meet?

Question 4: In order to discharge the battery, what conditions should we meet?

Question 5: In order to absorb reactive power from the AC system, what conditions should we meet?

Question 6: In order to generate reactive power to the AC system, what conditions should we meet?

# References

1. Peng, F. Z., Shen, M., & Qian, Z. (2005). Maximum boost control of the z-source inverter. *IEEE Transactions on Power Electronics, 20*(4), 833–838.
2. Wang, C., Li, Y., Peng, K., Hong, B., Wu, Z., & Sun, C. (2013). Coordinated optimal design of inverter controllers in a micro-grid with multiple distributed generation units. *IEEE Transactions on Power Systems, 28*(3), 2679–2687.
3. Li, Y., Gao, W., & Jiang, J. (2014). Stability analysis of microgrids with multiple der units and variable loads based on MPT. In *2014 IEEE PES General Meeting\ Conference & Exposition* (pp. 1–5). IEEE.
4. Ngamroo, I., & Karaipoom, T. (2014). Improving low-voltage ride-through performance and alleviating power fluctuation of DFIG wind turbine in DC microgrid by optimal SMES with fault current limiting function. *IEEE Transactions on Applied Superconductivity, 24*(5), 1–5.
5. Kasa, N., Ogawa, H., Iida, T., & Iwamoto, H. (1999). A transformer-less inverter using buck-boost type chopper circuit for photovoltaic power system. In *Proceedings of the IEEE 1999 International Conference on Power Electronics and Drive Systems. PEDS'99 (Cat. No. 99TH8475)* (Vol. 2, pp. 653–658). IEEE.
6. Noguchi, T., Togashi, S., & Nakamoto, R. (2002). Short-current pulse-based maximum-power-point tracking method for multiple photovoltaic-and-converter module system. *IEEE Transactions on Industrial electronics, 49*(1), 217–223.
7. Bodur, M., & Ermis, M. (1994). Maximum power point tracking for low power photovoltaic solar panels. In *Proceedings of MELECON'94. Mediterranean Electrotechnical Conference* (pp. 758–761). IEEE.
8. Zhao, H., Wu, Q. J., & Kawamura, A. (2004). An accurate approach of nonlinearity compensation for VSI inverter output voltage. *IEEE Transactions on Power Electronics, 19*(4), 1029–1035.
9. González, R., Lopez, J., Sanchis, P., & Marroyo, L. (2007). Transformerless inverter for single-phase photovoltaic systems. *IEEE Transactions on Power Electronics, 22*(2), 693–697.
10. Chen, Y., & Smedley, K. (2008). Three-phase boost-type grid-connected inverter. *IEEE Transactions on Power Electronics, 23*(5), 2301–2309.

11. Mantooth, H. A., & Hefner, A. R. (1997). Electrothermal simulation of an IGBT PWM inverter. *IEEE Transactions on Power Electronics, 12*(3), pp. 474–484.
12. Cho, K. M., Oh, W. S., Kim, Y. T., & Kim, H. J. (2007). A new switching strategy for pulse width modulation (PWM) power converters. *IEEE Transactions on Industrial Electronics, 54*(1), 330–337.
13. Holtz, J. (1992). Pulsewidth modulation-a survey. *IEEE Transactions on Industrial Electronics, 39*(5), 410–420.
14. Mukherjee, R., Patra, A., & Banerjee, S. (2009). Impact of a frequency modulated pulsewidth modulation (PWM) switching converter on the input power system quality. *IEEE Transactions on Power Electronics, 25*(6), 1450–1459.
15. Da Silva, E. R. C., dos Santos, E. C., & Jacobina, B. (2011). Pulsewidth modulation strategies. *IEEE Industrial Electronics Magazine, 5*(2), 37–45.
16. Zhou, K., & Wang, D. (2002). Relationship between space-vector modulation and three-phase carrier-based PWM: a comprehensive analysis [three-phase inverters]. *IEEE Transactions on Industrial Electronics, 49*(1), 186–196.
17. Liu, H. L., & Cho, G. H. (1994). Three-level space vector PWM in low index modulation region avoiding narrow pulse problem. *IEEE Transactions on Power Electronics, 9*(5), 481–486.
18. Nguyen, M.-K., & Choi, Y.-O. (2017). PWM control scheme for quasi-switched-boost inverter to improve modulation index. *IEEE Transactions on Power Electronics, 33*(5), 4037–4044.
19. Joos, G., & Espinoza, J. R. (1999). Three-phase series VAR compensation based on a voltage-controlled current source inverter with supplemental modulation index control. *IEEE Transactions on Power Electronics, 14*(3), 587–598.
20. Shen, D., & Lehn, P. (2002). Modeling, analysis, and control of a current source inverter-based statcom. *IEEE Transactions on Power Delivery, 17*(1), 248–253.
21. Schonardie, M. F., & Martins, D. C. (2008). Three-phase grid-connected photovoltaic system with active and reactive power control using dq0 transformation. In *2008 IEEE Power Electronics Specialists Conference* (pp. 1202–1207). IEEE.
22. Baimel, D., Belikov, J., Guerrero, J. M., & Levron, Y. (2017). Dynamic modeling of networks, microgrids, and renewable sources in the dq0 reference frame: A survey. *IEEE Access, 5*, 21323–21335.
23. Bimbhra, P., & Kaur, S. (2012). *Power electronics* (Vol. 2). Khanna Publishers.
24. Kazmierkowski, M. P., Krishnan, R., & Blaabjerg, F. (2002). *Control in power electronics*. Elsevier.
25. Mohammadi, P., & Mehraeen, S. (2016). Challenges of PV integration in low-voltage secondary networks. *IEEE Transactions on Power Delivery, 32*(1), 525–535.
26. Seguin, R., Woyak, J., Costyk, D., Hambrick, J., & Mather, B. (2016). *High-penetration pv integration handbook for distribution engineers*. National Renewable Energy Lab. (NREL), Golden, CO (United States), Tech. Rep.
27. Hughes, F. M., Anaya-Lara, O., Jenkins, N., & Strbac, G. (2005). Control of DFIG-based wind generation for power network support. *IEEE Transactions on Power Systems, 20*(4), 1958–1966.
28. Li, S., Haskew, T. A., Williams, K. A., & Swatloski, R. P. (2011). Control of DFIG wind turbine with direct-current vector control configuration. *IEEE transactions on Sustainable Energy, 3*(1), 1–11.

# Chapter 7
# Control of Microgrids

## 7.1 Microgrids

Power electricity has become a crucial and indispensable part of our daily life to promote the growth and development of the society. The conventional top-down monopoly energy configuration is dominated by large-scale fossil and nuclear power plants, which is hard to meet the increasing need of sustainable development of energy, as their emission is detrimental to the environment [1, 2]. Moreover, this traditional centralized energy configuration which unidirectionally sends electricity to passive consumers is aged and vulnerable to extreme or cascading conditions [3, 4], e.g., hurricane, earthquake, equipment failure, or cybercrime. For instance, a major blackout hit New York Manhattan on July 13, 2019, which was attributed to a 13 kV cable that burned up at West 64th Street and West End Avenue. This power outage quickly spread out and plunged a broad swath of Manhattan into darkness for up to 5 h on Saturday night, affecting over 73,000 customers and causing tens of millions of dollars' economic loss. All of these issues deeply block the contribution of power grid to the sustainable development of the whole society. Since sustainability has become the centre of recent national policies, strategies and development plans of many countries, microgrids have been deployed in recent years to seek an edge toward energy sustainability.

A microgrid is defined as a group of interconnected loads and distributed energy resources (DERs) within clearly defined electrical boundaries that acts as a single controllable entity with respect to the grid and can connect and disconnect from the grid to enable it to operate in both grid-connected or island-mode [4, 5]. A microgrid is illustrated in Fig. 7.1. CIGRÉ C6.22 Working Group (Microgrid Evolution Roadmap) also defines microgrids as electricity distribution systems containing loads and distributed energy resources, (such as distributed generators, storage devices, or controllable loads) that can be operated in a controlled, coordinated way either connected to the main power network or while islanded [6–9].

Microgrid has been recognized as a promising archetype to enhance the operations of low- or medium- voltage distribution networks and facilitate the high

© Springer Nature Switzerland AG 2022

Y. Li, *Cyber-Physical Microgrids*, https://doi.org/10.1007/978-3-030-80724-5_7

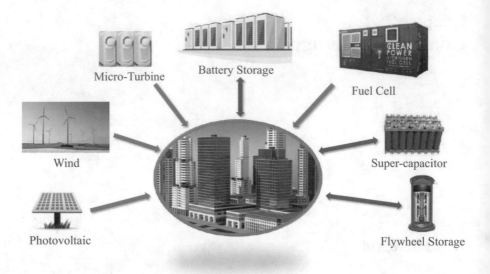

**Fig. 7.1** Microgrid system

penetration of DERs, such as wind and photovoltaic [10, 11]. As an effective means of integrating DERs into power systems, microgrid is able to offer consistent, flexible, affordable, reliable, and resilient local energy generation and delivery [12–14].

## 7.2   Hierarchical Control

To realize the above functions of microgrids, control system is playing an important role. The three-layer hierarchical control scheme is usually adopted to control the microgrid system [15–18], as shown in Fig. 7.2.

*Primary Control* The primary control is mostly used for the island microgrid systems [16, 19, 20]. The goal of the primary control is to stabilize the voltage, current, and frequency of the island system, and meanwhile, reaching a proper power sharing among DERs. So, decentralized control theory is usually adopted. The timescale of this control is second level.

*Secondary Control* The secondary control can be used either in the island mode or grid-connected mode [17]. For the island microgrids, the goal of the secondary control is to restore system's frequency and voltage. For the grid-connected mode, the goal is to control the exchange power of the Point of Common Coupling (PCC) between the microgrid and the bulk system. The secondary control can be also used

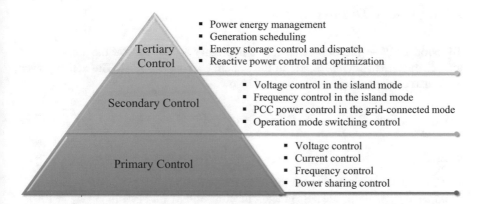

**Fig. 7.2** Three-layer hierarchical control scheme for microgrids

to switch the microgrid system between different operational modes. The timescale of this control is minute level.

*Tertiary Control* The tertiary control is mostly used for managing the electric energy in the microgrids [16, 17, 21], such as scheduling power generations of dispatchable DERs, managing the charging and discharging of energy storage units, controlling reactive power, etc. The timescale of this control is hours or days.

In this book, examples of primary control and secondary control are mainly introduced.

## 7.3 Droop Control

Droop control is one of the most popular types of primary control that regulates the frequency or voltage by adjusting the output active power or reactive power of a controlled asset [19]. So, droop control can be applied to battery, flywheel, supercapacitor, microturbine, etc. There are several types of droop controls, e.g., $P - f$ droop control $Q - V$ droop control, $P - V$ droop control, $Q - f$ droop control, etc.

Since the inverter plays an important role in integrating those dispatchable DERs into the AC system, droop control is mostly implemented in the control system of the inverter. According to (6.74) and (6.75), at the steady state, the voltages $V_I$ and $V_S$ should be around 1.0 p.u., so the inverter's active power output is mainly determined by the angle difference $\beta - \alpha$. The voltage difference mainly impacts the inverter's reactive power output. So, here we introduce $P - f$ droop control and $Q - V$ droop control.

### 7.3.1   *P − f Droop Control*

To introduce $P − f$ droop control, it is necessary to first look at the active power load characteristics with respect to the system's frequency [22]. The active power load characteristics can be summarized below.

$$P_{load} = a_0 + a_1 \cdot P_n\left(\frac{f}{f_n}\right) + a_2 \cdot P_n\left(\frac{f}{f_n}\right)^2 + a_3 \cdot P_n\left(\frac{f}{f_n}\right)^3 + \cdots + a_k \cdot P_n\left(\frac{f}{f_n}\right)^k, \quad (7.1)$$

where,

- $P_n$ is the active power load at the rated frequency.
- $a_0, a_1, a_2, a_3, \cdots a_k$ are percentages, meeting $a_0 + a_1 + a_2 + a_3 + \cdots + a_k = 100\%$.
- $f$ is the system's frequency.
- $f_n$ is the system's rated frequency.

Considering the microgrid system is only allowed to operate at a small range of frequency deviation from its rated value, we can only consider the linear part of the active power load characteristics within this range, so the load characteristics can be simplified in (7.2).

$$P_{load} = a_0 + a_1 \cdot P_n\left(\frac{f}{f_n}\right) \quad (7.2)$$

The major idea of $P − f$ droop control is that DERs will increase their active power outputs through losing the nominal frequency a bit. The idea is illustrated in Fig. 7.3.

Based on the active power load characteristics, the changes of DER active power output shown in Fig. 7.3 is analyzed below.

- Assume at the beginning the DER operates at the point A, where the frequency is rated value $f_n$ and the active power output of DER is $P_A$. In this scenario, the active power load is equal to the DER power output.
- When the active power load increases, the DER active power output is also expected to increase to keep the power balance, such as the point B. In this

**Fig. 7.3** Illustration of $P − f$ droop control

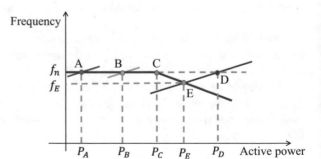

scenario, the DER will increase its active power output to $P_B$, and meanwhile, the frequency is still able to maintain at the rated value.

- When we keep the rated frequency, the DER's active power output limit is $P_C$. Beyond this limit, DER could not maintain the rated frequency when keeping the power balance.
- When the active power load increases to $P_D$, we expect the DER to increase its active power generation for balancing the power. However, it is beyond the capability of the DER. Therefore, one solution is that the DER generates more power, but the system's frequency is reduced a bit. According to (7.2), the decrease of frequency will lead to a decrease of active power load. Eventually, the DER power generation and load are balanced at the point E, namely, the system's frequency is $f_E$ and DER generates $P_E$ active power.

Based on the above analysis, the $P - f$ droop control can be expressed as,

$$f = f_n - K_P \Delta P \qquad (7.3)$$

where,

- $f$ is the island microgrid system's frequency to be determined.
- $K_P$ is the active power droop coefficient.
- $f_n$ is the system's rated frequency.
- $\Delta P$ is the active power increment.

## 7.3.2 $Q - V$ Droop Control

To introduce $Q - V$ droop control, it is necessary to look at the reactive power load characteristics with respect to its bus voltage [22]. The reactive power load characteristics can be summarized below.

$$Q_{load} = b_0 + b_1 \cdot Q_n\left(\frac{V}{V_n}\right) + b_2 \cdot Q_n\left(\frac{V}{V_n}\right)^2 + b_3 \cdot Q_n\left(\frac{V}{V_n}\right)^3 + \cdots + b_k \cdot Q_n\left(\frac{V}{V_n}\right)^k,$$
$$(7.4)$$

where,

- $Q_n$ is the reactive power load at the rated bus voltage.
- $b_0, b_1, b_2, b_3, \cdots b_k$ are percentages, meeting $b_0 + b_1 + b_2 + b_3 + \cdots + b_k = 100\%$.
- $V$ is the power load's bus voltage.
- $V_n$ is the rated value of the power load's bus voltage.

Considering the microgrid system is only allowed to operate at a small range of voltage deviation from its rated value, we can only consider the linear part of the active power load characteristics within this range, so the load characteristics can be simplified in (7.5).

**Fig. 7.4** Illustration of $Q - V$ droop control

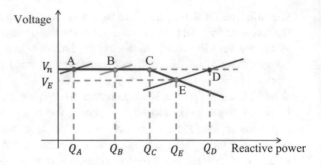

$$Q_{load} = b_0 + b_1 \cdot Q_n\left(\frac{V}{V_n}\right) \tag{7.5}$$

The major idea of $Q - V$ droop control is that DERs will increase their reactive power outputs through reducing the bus voltage a bit. The idea is illustrated in Fig. 7.4.

Based on the reactive power load characteristics, the changes of DER reactive power output shown in Fig. 7.4 is analyzed below.

- Assume at the beginning the DER operates at the point A, where the DER's bus voltage is the value $V_n$ and the reactive power output of DER is $Q_A$. In this scenario, the reactive power load is equal to the DER power output.
- When the reactive power load increases, the DER reactive power output is also expected to increase to keep the power balance, such as the point B. In this scenario, the DER will increase its reactive power output to $Q_B$, and meanwhile, the we can still maintain the bus voltage at the rated value.
- When we keep the rated bus voltage, the DER's reactive power output limit is $Q_C$. Beyond this limit, DER could not maintain the rated bus voltage when keeping the power balance.
- When the reactive power load increases to $Q_D$, we expect the DER to increase its reactive power generation for balancing the power. However, it is beyond the capability of the DER. Therefore, one solution is that the DER generates more power, but the bus voltage is reduced a bit. According to (7.5), the decrease of bus voltage will also lead to a decrease of the reactive power load. Eventually, the DER power generation and load are balanced at the point E, namely, the bus voltage is $V_E$ and DER generates $Q_E$ reactive power.

Based on the above analysis, the $Q - V$ droop control can be expressed as,

$$V = V_n - K_Q \Delta Q \tag{7.6}$$

where,

- $V$ is the DER's bus voltage to be determined.
- $K_Q$ is the reactive power droop coefficient.

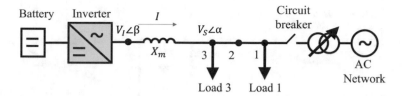

**Fig. 7.5**  Battery integration for Example 7.1

- $V_n$ is the rated value of the DER's bus voltage.
- $\Delta Q$ is the reactive power increment.

**Example 7.1**
A battery is integrated into the AC system through an inverter, as shown in Fig. 7.5. When the circuit breaker is open, the inverter is controlled under droop strategy. The droop coefficients is 0.01 for the active power and 0.002 for the reactive power. The original active power load at 60 Hz is 20 kW and reactive power is 10 kVAR. The original voltage at bus 3 is 0.4 kV.
Questions:
When the total power load increases to 30 kW and 15 kVAR, what are the system frequency and bus 3 voltage?

Solution:
According to (7.3), we can have the following equation,

$$f = f_n - K_P \Delta P = 60 - 0.01 \times (30 - 20) = 59.9 (Hz) \tag{7.7}$$

According to (7.6), we can have the following equation,

$$V = V_n - K_Q \Delta Q = 0.4 - 0.002 \times (15 - 10) = 0.39 (kV) \tag{7.8}$$

So, we can see the system's frequency in the new operation point is 59.9Hz and the bus 3's voltage is 0.39 kV. From the above calculation and analysis, we can also see the global frequency is involved in the active power calculation; however, the local voltage is involved in the reactive power calculation. ∎

### 7.3.3  Droop Controls for Multiple DERs

The basic droop controls are introduced for individual DERs. When the microgrid has multiple DERs, it is important to figure out how the increment of power load is

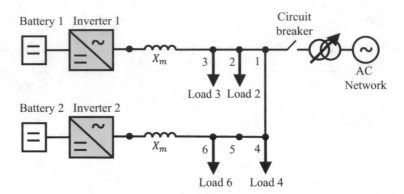

**Fig. 7.6**  System topology for Example 7.2

shared among those droop-controlled DERs. Here, we use Example 7.2 to analyze the droop control when the island microgrid system has multiple DERs with droop strategy.

**Example 7.2**

Two battery units are integrated into the AC system through inverters, as shown in Fig. 7.6. Before the circuit breaker is open, Battery 1 generates $P_{0,1}$ active power and Battery 2 generates $P_{0,2}$ active power. $P_{0,1}$ and $P_{0,2}$ are the power output limits that we can keep the system operating at the rated frequency. Before the circuit breaker is open, $\Delta P$ active power and $\Delta Q$ reactive power are delivered from the AC network to the power loads. After the circuit breaker is open, the two inverters are controlled under droop strategy. The active power droop coefficients are $K_{P1}$ for the Battery 1 and $K_{P2}$ for the Battery 2. The reactive power droop coefficients are $K_{Q1}$ for the Battery 1 and $K_{Q2}$ for the Battery 2.

Question 1: Assume the active power load parameter $a_0$ is equal to zero and $a_1$ is equal to $a$. After the circuit breaker is open, what is the system's frequency and how much active power are generated by Battery 1 and Battery 2, respectively?

Question 2: Assume the reactive power load parameter $b_0$ is equal to zero and $b_1$ is equal to $b$. After the circuit breaker is open, what is the batteries' bus voltages and how much reactive power are generated by Battery 1 and Battery 2, respectively?

Solution: Based on the droop control analysis for individual DERs, we can also analyze the two batteries' droop controls for their active power and reactive power outputs, respectively.

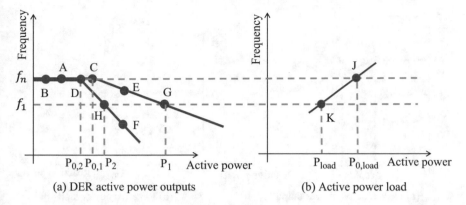

(a) DER active power outputs  (b) Active power load

**Fig. 7.7** Illustration of $P - f$ droop control for multiple DERs

*Active Power Droop Control*

- Before the circuit breaker is open, Battery 1 operates at the point C and Battery 2 operates at the point D, where the system's frequency is rated value $f_n$, as illustrated in Fig. 7.7. In this scenario, we have the following active power balance equation (7.9) when we ignore the power loss over the lines.

$$P_{0,1} + P_{0,2} + \Delta P = P_{0,load} \tag{7.9}$$

- When the circuit breaker is open, according to the individual DER's droop control, the two batteries will increase their active power outputs, and meanwhile the system's frequency will decrease. For instance, Battery 1 increases its power output to operates at point E, and Battery 2 operates at point F. The total power load will change along the characteristic line.
- Considering the frequency is a global variable, eventually, Battery 1 generates $P_1$ power output and Battery 2 generates $P_2$ power output, i.e., Battery 1 operates at point G, and Battery 2 operates at point H to ensure the same frequency. Therefore, we have (7.10) and (7.11), and they are supposed to equal.

$$f = f_n - K_{P1} \cdot (P_1 - P_{0,1}) \tag{7.10}$$

$$f = f_n - K_2 \cdot (P_2 - P_{0,2}) \tag{7.11}$$

where the new power outputs $P_1$ and $P_2$ meet the following power balance equation in the new operation point.

$$P_1 + P_2 = P_{load} = a \cdot P_{0,load}\left(\frac{f}{f_n}\right) \tag{7.12}$$

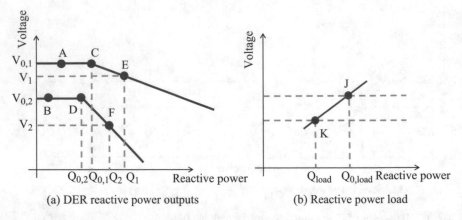

(a) DER reactive power outputs            (b) Reactive power load

**Fig. 7.8** Illustration of $Q - V$ droop control for multiple DERs

Solving (7.10), (7.11), and (7.12), we can get the island system's frequency and the two batteries' active power outputs.

*Reactive Power Droop Control*

- Before the circuit breaker is open, Battery 1 operates at the point C and Battery 2 operates at the point D, where the batteries generate $Q_{0,1}$ and $Q_{0,2}$ reactive power, respectively, as illustrated in Fig. 7.8. In this scenario, we have the following active power balance equation (7.13) when we ignore the power loss over the lines.

$$Q_{0,1} + Q_{0,2} + \Delta Q = Q_{0,load} \tag{7.13}$$

- When the circuit breaker is open, the two batteries will correspondingly adjust their reactive power outputs to keep the power balance. For instance, Battery 1 increases its power output to $Q_1$, operating at point E. And Battery 2 operates at point F, generating $Q_2$ reactive power. The decrease of bus voltage will lead to a decrease of reactive power loads.

$$V_3 = V_{3,0} - K_{Q1} \cdot (Q_1 - Q_{0,1}) \tag{7.14}$$

$$V_6 = V_{6,0} - K_{Q2} \cdot (Q_2 - Q_{0,2}) \tag{7.15}$$

where the new power outputs $Q_1$ and $Q_2$ meet the following power balance equation in the new operation point.

$$Q_{l,i} = b \cdot Q_{l,i0}\left(\frac{V_i}{V_{n,i}}\right) \tag{7.16}$$

where,

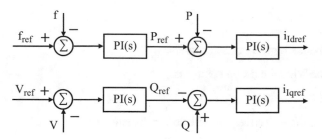

**Fig. 7.9** Example of $V - f$ Control for inverters

- $i = 2, 3, 4, 6$.
- $Q_{l,i0}$ is the original reactive power load at buses 2, 3, 4, and 6.
- $Q_{l,i}$ is the new reactive power load at buses 2, 3, 4, and 6.
- $V_i$ is bus voltage at node 2, 3, 4, and 6.
- $V_{n,i}$ is rated voltage at node 2, 3, 4, and 6.

Solving (7.14), (7.15), (7.16), and the power flow equation, we can get the bus voltages and the two batteries' reactive power outputs.

## 7.4   $V - f$ Control

Constant voltage and constant frequency control ($V - f$ control) is used to keep the voltage and frequency at the given value by adjusting the output active power or reactive power of DERs in the island microgrid system. Similar to droop control, there are several methods to implement $V - f$ control. One example is given in Fig. 7.9, where we keep the frequency at the nominal value through adjusting DER's active power output and control the voltage through adjusting DER's reactive power output.

## 7.5   DER Power Dispatch

DER power output dispatch is used to realize two goals. First, in the secondary control, for the grid-connected mode, DER power output dispatch is applied to control the exchange power of PCC between the microgrid and the bulk system. Second, in the tertiary control, DER power output dispatch is applied to manage the electric energy in the microgids. To realize those goals, a Microgrid Coordination Center (MCC) is usually used to adjust $P_{ref}$ and $Q_{ref}$ remotely via the communication network which will be introduced in Chap. 9. That is $\Delta_P$ and $\Delta_Q$ will be generated in the MCC and sent over the communication network to add to $P_{ref}$ and $Q_{ref}$.

## Problems

**7.1** Describe the concept of Microgrids.

**7.2** Describe what the droop control is.

**7.3** Describe what the $V - f$ control is.

**7.4** Describe the objectives of DER power dispatch.

**7.5** A battery is integrated into the AC system through an inverter, as shown in Fig. 7.5. When the circuit breaker is open, the inverter is controlled under droop strategy. The droop coefficients is 0.05 for the active power and 0.01 for the reactive power. The original active power load at 60 Hz is 50 kW and reactive power is 20 kVAR. The original voltage at bus 3 is 0.4 kV.

Question 1: When the total power load changes to 30 kW and 15 kVAR, what is the system's frequency?

Question 2: For the new power load, what is the bus 3's voltage?

**7.6** A battery is integrated into the AC system through an inverter, as shown in Fig. 7.5. When the circuit breaker is open, the inverter is controlled under droop strategy. The original active power load at 60 Hz is 30 kW and reactive power is 18 kVAR. The original voltage at bus 3 is 0.38 kV. When the power load increases to 42 kW and reactive power decreases to 15 kVAR, the system frequency becomes 59.5 Hz and bus 3 voltage is 0.40 kV.

Question 1: What is the droop coefficient for the active power control?

Question 2: What is the droop coefficient for the reactive power control?

**7.7** Three battery units are integrated into the AC system through inverters, as shown in Fig. 7.10. Before the circuit breaker is open, Battery 1 generates $P_{0,1}$ active power, Battery 2 generates $P_{0,2}$ active power, and Battery 3 generates $P_{0,3}$ active power. $P_{0,1}$, $P_{0,2}$, and $P_{0,3}$ are the power output limits that we can keep the system operating at the rated frequency. Before the circuit breaker is open, $\Delta P$ active power and $\Delta Q$ reactive power are delivered from the AC network to the power loads. After the circuit breaker is open, the three inverters are controlled under droop strategy. The active power droop coefficients are $K_{P1}$ for the Battery 1, $K_{P2}$ for the Battery 2, and $K_{P3}$ for the Battery 3. The reactive power droop coefficients are $K_{Q1}$ for the Battery 1, $K_{Q2}$ for the Battery 2, and $K_{Q3}$ for the Battery 3.

Question 1: Assume the active power load parameter $a_0$ is equal to zero and $a_1$ is equal to $a$. After the circuit breaker is open, what is the system's frequency?

Question 2: After the circuit breaker is open, how much active power are generated by Battery 1, Battery 2, and Battery 3, respectively?

Question 3: Assume the reactive power load parameter $b_0$ is equal to zero and $b_1$ is equal to $b$. After the circuit breaker is open, what is the batteries' bus voltages?

Question 4: After the circuit breaker is open, how much reactive power are generated by Battery 1, Battery 2, and Battery 3, respectively?

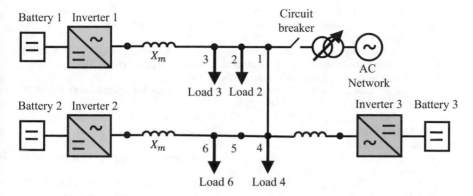

**Fig. 7.10** System topology for Problem 7.7

# References

1. Kumar, S., Katoria, D., & Sehgal, D. (2013). Environment impact assessment of thermal power plant for sustainable development. *International Journal of Environmental Engineering and Management, 4*(6), 567–572.
2. Twidell, J. (2006). *Renewable energy resources.* Routledge.
3. Anadon, L. D., GaLLaGher, K. S., & Bunn, M. G. (2009). DOE FY 2010 budget request and Recovery Act funding for energy research, development, demonstration, and deployment: Analysis and recommendations. Cambridge: Report for Energy Technology Innovation Policy research group, Belfer Center for Science and International Affairs, Harvard Kennedy School.
4. Bossart, S. (2012). DOE perspective on microgrids. In *Advanced Microgrid Concepts and Technologies Workshop.*
5. Li, Y., Zhang, P., & Luh, P. B. (2017). Formal analysis of networked microgrids dynamics. *IEEE Transactions on Power Systems, 33*(3), 3418–3427.
6. Marnay, C., Chatzivasileiadis, S., Abbey, C., Iravani, R., Joos, G., Lombardi, P., Mancarella, P., & von Appen, J. (2015). Microgrid evolution roadmap. In *2015 International Symposium on Smart Electric Distribution Systems and Technologies (EDST)* (pp. 139–144). IEEE.
7. Olivares, D. E., Mehrizi-Sani, A., Etemadi, A. H., Cañizares, C. A., Iravani, R., Kazerani, M., Hajimiragha, A. H., Gomis-Bellmunt, O., Saeedifard, M., Palma-Behnke, R., et al. (2014). Trends in microgrid control. *IEEE Transactions on Smart Grid, 5*(4), 1905–1919.
8. Lasseter, R. H., & Paigi, P. (2004). Microgrid: A conceptual solution. In *2004 IEEE 35th Annual Power Electronics Specialists Conference (IEEE Cat. No. 04CH37551)* (Vol. 6, pp. 4285–4290). IEEE.
9. Lasseter, R., Akhil, A., Marnay, C., Stephens, J., Dagle, J., Guttromson, R., Meliopoulous, A., Yinger, R., & Eto, J. (2002). The certs microgrid concept. *White paper for Transmission Reliability Program, Office of Power Technologies, US Department of Energy, 2*(3), 30.
10. Lasseter, R. H. (2007). CERTS microgrid. In *2007 IEEE International Conference on System of Systems Engineering* (pp. 1–5). IEEE.
11. Marnay, C., & Bailey, O. C. (2004). The CERTS microgrid and the future of the macrogrid. Lawrence Berkeley National Lab. (LBNL), Berkeley, CA (United States), Tech. Rep.
12. Nejabatkhah, F., & Li, Y. W. (2014). Overview of power management strategies of hybrid ac/dc microgrid. *IEEE Transactions on Power Electronics, 30*(12), 7072–7089.
13. Agrawal, P. (2006). Overview of DOE microgrid activities. In *Symposium on Microgrid, Montreal, June* (Vol. 23).
14. Piagi, P. (2005). *Microgrid control.* Ph.D. dissertation, University of Wisconsin–Madison.

15. Guerrero, J. M., Chandorkar, M., Lee, T.-L., & Loh, P. C. (2012). Advanced control architectures for intelligent microgrids–part I: Decentralized and hierarchical control. *IEEE Transactions on Industrial Electronics, 60*(4), 1254–1262.
16. Guerrero, J. M., Vasquez, J. C., Matas, J., De Vicuña, L. G., & Castilla, M. (2010). Hierarchical control of droop-controlled ac and dc microgrids–a general approach toward standardization. *IEEE Transactions on Industrial Electronics, 58*(1), 158–172.
17. Bidram, A., & Davoudi, A. (2012). Hierarchical structure of microgrids control system. *IEEE Transactions on Smart Grid, 3*(4), 1963–1976.
18. Vasquez, J. C., Guerrero, J. M., Miret, J., Castilla, M., & De Vicuna, L. G. (2010). Hierarchical control of intelligent microgrids. *IEEE Industrial Electronics Magazine, 4*(4), 23–29.
19. Wang, C., Li, Y., Peng, K., Hong, B., Wu, Z., & Sun, C. (2013). Coordinated optimal design of inverter controllers in a micro-grid with multiple distributed generation units. *IEEE Transactions on Power Systems, 28*(3), 2679–2687.
20. Vandoorn, T. L., Vasquez, J. C., De Kooning, J., Guerrero, J. M., & Vandevelde, L. (2013). Microgrids: Hierarchical control and an overview of the control and reserve management strategies. *IEEE Industrial Electronics Magazine, 7*(4), 42–55.
21. Babazadeh, H., Gao, W., Wu, Z., & Li, Y. (2013). Optimal energy management of wind power generation system in islanded microgrid system. In *2013 North American Power Symposium (NAPS)* (pp. 1–5). IEEE.
22. Martí, J. R., Ahmadi, H., & Bashualdo, L. (2013). Linear power-flow formulation based on a voltage-dependent load model. *IEEE Transactions on Power Delivery, 28*(3), 1682–1690.

# Chapter 8
# Modeling and Stability Analysis of Microgrids

## 8.1 Dynamic Modeling of Microgrids

Based on the control strategies and modeling of DERs, a microgrid system can be modeled by a set of differential-algebraic equations (DAEs) as given in (8.1), when the power electronic interfaces of DER units are modeled by using the average models [1–5], as introduced in Chap. 6.

$$\begin{cases} \dot{x} = f(x, y) \\ 0 = g(x, y), \end{cases} \qquad (8.1)$$

where,

- $x \in \mathbb{R}^n$ is the vector of state variables.
- $y \in \mathbb{R}^m$ is the vector of bus voltages and angles.

A typical microgrid system given in Fig. 8.1 is used as an example to demonstrate the dynamic operations of microgrid [6, 7]. The detailed impedance, power loads, and DER generations of the test system are provided in Tables 8.1, 8.2, and 8.3, respectively.

### 8.1.1 $V - f$ Control Time-Domain Simulations

Time domain simulations are implemented to demonstrate the transient dynamics of the microgrid. In Fig. 8.1, the Circuit Breaker 0 is open, so the microgrid operates in the island mode. Micro-turbine 13 is under $V - f$ control to maintain the system's frequency and meanwhile keeping bus 13's voltage at the given value 1.0053 p.u. All the PVs are under MPPT control. All the batteries and fuel cells are under constant power control, i.e., the references of their power outputs do not change.

© Springer Nature Switzerland AG 2022
Y. Li, *Cyber-Physical Microgrids*, https://doi.org/10.1007/978-3-030-80724-5_8

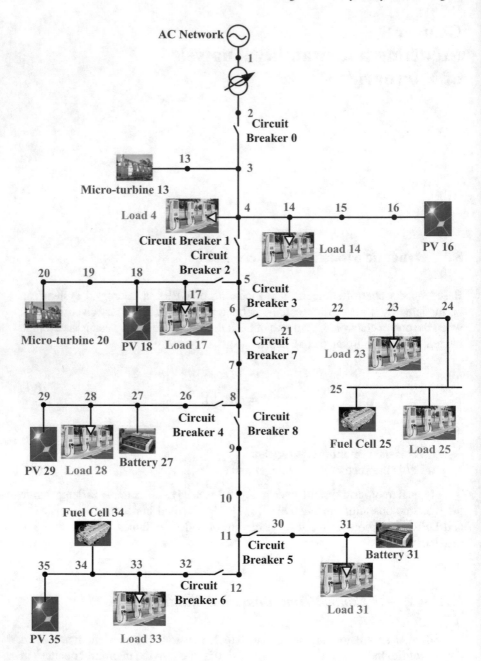

**Fig. 8.1** A typical microgrid test system

**Table 8.1** Line impedances between buses in Fig. 8.1

| From | To | $R(\Omega/\text{km})$ | $L$ (H/km) | Length (m) |
|------|-----|------------------------|-------------------------|------------|
| 3 | 4 | 0.2712 | $0.1856 \times 10^{-3}$ | 45 |
| 3 | 13 | 3.1200 | $0.2308 \times 10^{-3}$ | 50 |
| 4 | 14 | 3.1200 | $0.2308 \times 10^{-3}$ | 50 |
| 14 | 15 | 3.1200 | $0.2308 \times 10^{-3}$ | 50 |
| 15 | 16 | 3.1200 | $0.2308 \times 10^{-3}$ | 50 |
| 5 | 6 | 0.2712 | $0.1856 \times 10^{-3}$ | 45 |
| 17 | 18 | 2.4591 | $0.3256 \times 10^{-3}$ | 20 |
| 18 | 19 | 2.4591 | $0.3256 \times 10^{-3}$ | 20 |
| 19 | 20 | 2.4591 | $0.3256 \times 10^{-3}$ | 20 |
| 21 | 22 | 0.5260 | $0.3025 \times 10^{-3}$ | 30 |
| 22 | 23 | 0.5260 | $0.3025 \times 10^{-3}$ | 30 |
| 23 | 24 | 0.5260 | $0.3025 \times 10^{-3}$ | 30 |
| 24 | 25 | 0.5260 | $0.3025 \times 10^{-3}$ | 30 |
| 7 | 8 | 0.2712 | $0.1856 \times 10^{-3}$ | 45 |
| 26 | 27 | 0.7849 | $0.1906 \times 10^{-3}$ | 40 |
| 27 | 28 | 0.7849 | $0.1906 \times 10^{-3}$ | 40 |
| 28 | 29 | 0.7849 | $0.1906 \times 10^{-3}$ | 40 |
| 9 | 10 | 0.2712 | $0.1856 \times 10^{-3}$ | 45 |
| 10 | 11 | 0.2712 | $0.1856 \times 10^{-3}$ | 45 |
| 30 | 31 | 4.5600 | $0.3026 \times 10^{-3}$ | 30 |
| 11 | 12 | 0.2712 | $0.1856 \times 10^{-3}$ | 45 |
| 32 | 33 | 1.2613 | $0.1956 \times 10^{-3}$ | 30 |
| 33 | 34 | 1.2613 | $0.1956 \times 10^{-3}$ | 30 |
| 34 | 35 | 1.2613 | $0.1956 \times 10^{-3}$ | 30 |

**Table 8.2** Power loads at each bus in Fig. 8.1

| Bus | $P_n$ (kW) | $Q_n$ (kVAR) | Bus | $P_n$ (kW) | $Q_n$ (kVAR) |
|-----|-----------|--------------|-----|-----------|--------------|
| 4 | 32.69 | 15.97 | 25 | 60.54 | 40.63 |
| 14 | 74.69 | 41.26 | 28 | 61.35 | 37.59 |
| 17 | 41.56 | 27.64 | 31 | 58.21 | 36.78 |
| 23 | 59.63 | 38.57 | 33 | 65.31 | 39.45 |

**Table 8.3** DER generations at each bus in Fig. 8.1

| Bus | $P_n$ (kW) | $Q_n$ (kVAR) | Bus | $P_n$ (kW) | $Q_n$ (kVAR) |
|-----|-----------|--------------|-----|-----------|--------------|
| 13 | 5.36 | 3.65 | 27 | 25.69 | 15.75 |
| 16 | 15.36 | 10.25 | 29 | 25.00 | 12.00 |
| 18 | 20.00 | 15.25 | 31 | 45.30 | 30.65 |
| 20 | 10.25 | 5.68 | 34 | 36.78 | 14.23 |
| 25 | 64.59 | 54.26 | 35 | 29.24 | 14.96 |

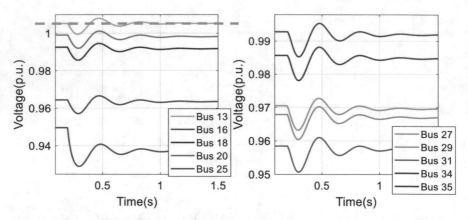

**Fig. 8.2** Voltage changes of DER buses when the power load changes in the Case 1 of $V - f$ control scenario

All the power loads are treated as constant power loads. The test system is stable initially. Three cases are carried out to show the system's dynamics and introduced as follows.

*Case 1* At 0.2 s, a very small power load disturbance is introduced, i.e., the Load 14's active power increases by 0.01%. Figure 8.2 shows the voltage changes of DER buses when the power load changes. From Fig. 8.2, we can see,

- With the $V - f$ control strategy, bus 13 can keep its voltage at the given reference after being stable again, i.e., 1.0053 p.u.
- The voltages of other buses correspondingly change. In particular, there is a decrease of the voltage of bus 16 because bus 16 is very close to bus 14, which connects to the changing power load.

*Case 2* At 0.2 s, a relatively large power load disturbance is considered, i.e., the Load 14's active power increases by 100%. Figure 8.3 shows the DER bus voltages' changes when the power load changes. Findings are summarized below.

- Bus 13's voltage in Fig. 8.2 shows it can be kept at the given reference after the system being stable again, i.e., 1.0053 p.u.
- The voltages of other buses correspondingly change. In particular, there is a big decrease of bus 16's voltage due to the small electrical distance between bus 16 and bus 14.
- It also demonstrates the large disturbance can induce more severe system dynamics because it apparently takes more time for the system to be stable again.

*Case 3* At 0.2 s, a very large power load disturbance is introduced, i.e., the Load 14's active power increases by 152%. Figure 8.4 shows the DER bus voltage changes, from which we can see the microgrid system is going through very severe

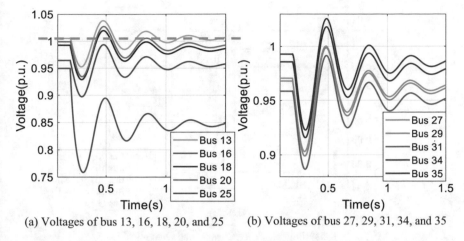

**Fig. 8.3** Voltage changes of DER buses when the power load changes in the Case 2 of $V - f$ control scenario

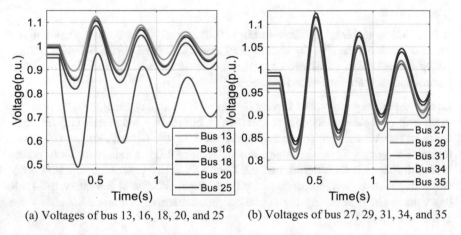

**Fig. 8.4** Voltage changes of DER buses when the power load changes in the Case 3 of $V - f$ control scenario

dynamics. Further tests also show this power load increment is a critical operation point, beyond which the system will be unstable.

## 8.1.2 Droop Control Time-Domain Simulations

Droop control is a critical and desired function of microgrids, especially for the island operations. Time domain simulations on Fig. 8.1 are implemented to demonstrate the transient dynamics of the microgrid when DERs are under droop

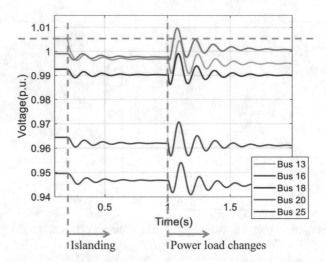

**Fig. 8.5** Voltage changes of DER buses in the Case 1 of droop control scenario

control. Micro-turbine 13, Micro-turbine 20, Battery 27, and Battery 31, are under droop control. All the PVs are under MPPT control. Other DERs are under constant power control. Two cases are carried out to demonstrate the function of droop control and they are introduced below.

*Case 1*   At the beginning, the Circuit Breaker 0 is closed, so the microgrid operates in the grid-connected mode. More specifically, both active and reactive power flows are from the AC network to the microgrid. At 0.2 s, the Circuit Breaker 0 is open, turning the microgrid to the island mode. Then at 1.0 s, a relatively large power load disturbance is considered, i.e., the Load 17's active power increases by 50%, while its reactive power decreases by 50%. Figures 8.5 and 8.6 show the DER bus voltage changes during the above operational scenario. Figure 8.7 shows the system's frequency.

From Figs. 8.5, 8.6, and 8.7, we can see that,

- When the microgrid is in the grid-connected mode, its frequency can be kept at 60 Hz, which is determined by the AC network.
- Before the circuit breaker is open, the power flow is from the AC network to the microgrid, which means there is a power deficit in the microgrid. So, after the circuit breaker is open, according to the droop control principle, DERs are expected to generate more active and reactive power to keep the balance. Correspondingly, the system's frequency and bus voltages will be reduced, which can be seen from the results at 0.2 s.
- When the active and reactive power loads have different changes, e.g., increasing active power loads and decreasing reactive power loads, some of the bus voltages may increase (e.g. buses 13 and 31) and others may decrease (e.g., bus 16). It

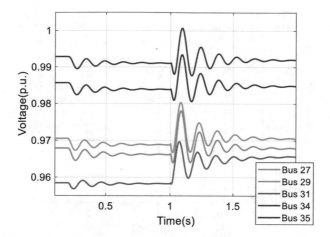

**Fig. 8.6** DER voltage changes in the Case 1 of droop control scenario

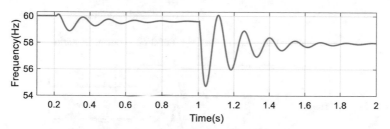

**Fig. 8.7** Frequency changes in the Case 1 of droop control scenario

indicates the active and reactive power are still coupled even though we control them separately.

*Case 2* At the beginning, the Circuit Breaker 0 is open, so the microgrid is initially operating in the island mode. At 0.2 s, a power load disturbance is considered, i.e., the Load 17's active and reactive power increase by 10%. Then at 1.0 s, the Load 17's active and reactive power decrease by 20%. Figures 8.8 and 8.9 show the DER bus voltage changes during the above operational scenario. Figures 8.10 and 8.11 show the active and reactive power outputs of droop-controlled DERs. Figure 8.12 shows the system's frequency. From the results, we can see that,

- When the active power loads increases in the island microgrid, the system's frequency will decrease and vice versa. Correspondingly, the droop-controlled DERs increase their active power output and vice versa.
- When the reactive power loads increases in the island microgrid, in order to generate more reactive power from a DER, its bus voltage will decrease and vice

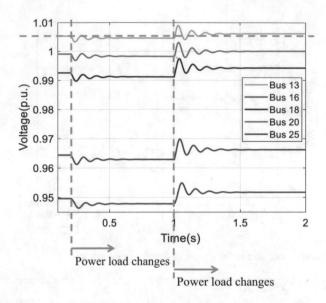

**Fig. 8.8**  Voltage changes of DER buses in the Case 2 of droop control scenario

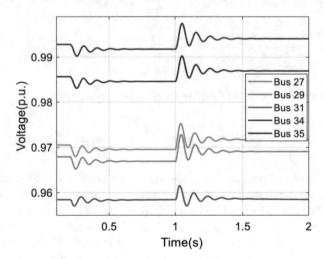

**Fig. 8.9**  DER voltage changes in the Case 2 of droop control scenario

versa. Correspondingly, the droop-controlled DERs increase their reactive power output and vice versa.

• Under proper droop control, DERs can adjust their active and reactive power outputs for reaching a new stable operational point.

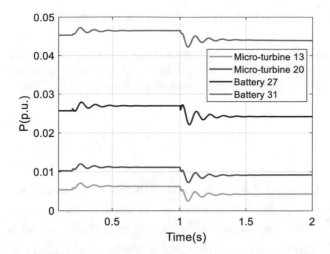

**Fig. 8.10**  Changes of DER active power outputs in the Case 2 of droop control scenario

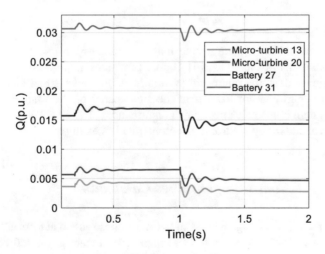

**Fig. 8.11**  Changes of DER reactive power outputs in the Case 2 of droop control scenario

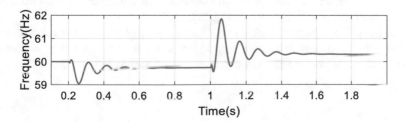

**Fig. 8.12**  Changes of frequency in the Case 2 of droop control scenario

## 8.2   Small-Signal Stability Analysis of Microgrids

Small-signal stability analysis investigates the system's stability subject to a relative small disturbance [8–12]. Based on (8.1), linearizing the nonlinear microgrid system at one operational point, we can obtain the following model, when $g_y$ is non-singular.

$$\Delta \dot{x} = \left( f_x - f_y g_y^{-1} g_x \right) \Delta x, \tag{8.2}$$

where,

- $f_x = \frac{\partial f}{\partial x}$ is the partial derivative matrix with respect to the state variables.
- $g_x = \frac{\partial g}{\partial x}$ is the partial derivative matrix with respect to the state variables.
- $f_y = \frac{\partial f}{\partial y}$ is the partial derivative matrix with respect to the algebraic variables.
- $g_y = \frac{\partial g}{\partial y}$ is the partial derivative matrix with respect to the algebraic variables.

Therefore, based on (8.2), the system's state matrix $A$ is defined as follows.

$$A = f_x - f_y g_y^{-1} g_x \tag{8.3}$$

The eigenvalues of the system's state matrix $A$ determine the system's stability subject to small disturbances [13, 14]. If all eigenvalues are located in the left panel of the imaginary axis, the microgrid system is stable under small disturbances. If at least one eigenvalue are located in the right panel of the imaginary axis, the microgrid system is unstable under small disturbances. The eigenvalues of the above $V - f$ control and droop control are demonstrated as examples.

### 8.2.1   $V - f$ Control Eigenvalues Calculation

In the $V - f$ control, all eigenvalues of the microgrid system at 0.0 s are calculated and shown in Fig. 8.13. We can see all eigenvalues locate in the left panel of the imaginary axis, which indicates the linearized $V - f$ controlled microgrid system is stable. Figure 8.14 shows the zoom-in part of the eigenvalues. The highlighted two pairs of eigenvalues in Fig. 8.14 mainly determine the dynamics of the microgrid system under disturbances shown in Fig. 8.2.

Studies on the state matrix $A$ show it has a zero eigenvalue. It is caused by the fact that relative voltage angles are used in the power flow computation, as absolute angles make little engineering sense and are hard to define in practice.

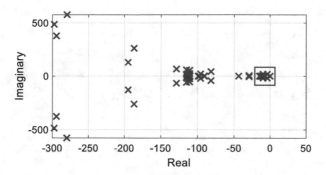

**Fig. 8.13** All eigenvalues of the microgrid test system in the $V - f$ control

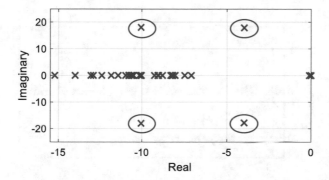

**Fig. 8.14** Two pairs of eigenvalues that mainly determine the dynamics of the microgrid

### 8.2.2 Droop Control Eigenvalues Calculation

In the droop control, all eigenvalues of the microgrid system at 0.9 s are calculated and shown in Fig. 8.15. We can see all eigenvalues locate in the left panel of the imaginary axis, which indicates the linearized droop controlled microgrid system is stable. Figure 8.16 shows the zoom-in part of the eigenvalues. From the zoom-in eigenvalue distributions, we can see it is very different from the eigenvalues of $V - f$ control. The highlighted pair of eigenvalues in Fig. 8.16 mainly determine the dynamics of the droop-controlled microgrid system.

## Problems

**8.1** Describe the dynamic modeling of microgrids.

**8.2** What is the small-signal stability of the microgrid systems?

**8.3** Describe how the eigenvalues determine the stability of the microgrid system.

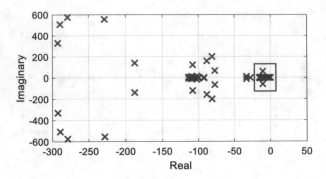

**Fig. 8.15** All eigenvalues of the microgrid test system in the droop control

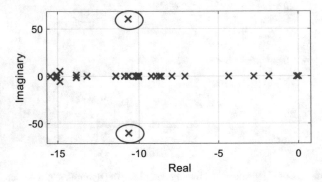

**Fig. 8.16** One pair of eigenvalues that mainly determine the dynamics of droop-controlled microgrid

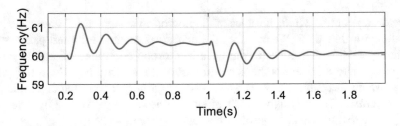

**Fig. 8.17** The microgrid system's frequency for Problem 8.4

**8.4** In the microgrid given in Fig. 8.1, the system is initially connecting to the AC network. Then at 0.2 s, the Circuit Breaker 0 is open. After that at 1.0 s, the power loads in the island system change. The microgrid system's frequency is given in Fig. 8.17.

Question 1: In the above scenario, before the Circuit Breaker 0 is open, is the power exchange from AC network to the microgrid or inverse?

Question 2: After the Circuit Breaker 0 is open, does the droop-controlled DERs increase or decrease their active power outputs?

Question 3: Compared to 0.9 s, does the power load increase or decrease?

# References

1. Li, Y., Zhang, P., & Luh, P. B. (2017). Formal analysis of networked microgrids dynamics. *IEEE Transactions on Power Systems, 33*(3), 3418–3427.
2. Li, Y., Zhang, P., Althoff, M., & Yue, M. (2018). Distributed formal analysis for power networks with deep integration of distributed energy resources. *IEEE Transactions on Power Systems, 34*(6), 5147–5156.
3. Li, Y., Zhang, P., & Yue, M. (2018). Networked microgrid stability through distributed formal analysis. *Applied Energy, 228*, 279–288.
4. Li, Y., Huang, D., Zhang, Y., & Orekan, T. (2021). Resilience analysis of cyber-physical networked microgrids with communication latency. In *2021 IEEE Power & Energy Society General Meeting (PESGM)* (pp. 1–5). IEEE.
5. Jiang, X., Li, Y., Du, L., & Huang, D. (2021). Identifying HOPF bifurcations of networked microgrids induced by the integration of EV charging stations. In *2021 IEEE Transportation Electrification Conference & Expo (ITEC)* (pp. 1–5). IEEE.
6. Li, Y., Qin, Y., Zhang, P., & Herzberg, A. (2018). SDN-enabled cyber-physical security in networked microgrids. *IEEE Transactions on Sustainable Energy, 10*(3), 1613–1622.
7. Li, Y., & Du, L. (2021). Programmable and reconfigurable cyber-physical networked microgrids through software-defined networking. In *2021 IEEE Transportation Electrification Conference & Expo (ITEC)* (pp. 1–5). IEEE.
8. Mondal, D., Chakrabarti, A., & Sengupta, A. (2020). *Power system small signal stability analysis and control.* Academic Press.
9. Li, Y., Zhang, P., Ren, L., & Orekan, T. (2016). A Geršgorin theory for robust microgrid stability analysis. In *2016 IEEE Power and Energy Society General Meeting (PESGM)* (pp. 1–5). IEEE.
10. Wang, C., Li, Y., Peng, K., Hong, B., Wu, Z., & Sun, C. (2013). Coordinated optimal design of inverter controllers in a micro-grid with multiple distributed generation units. *IEEE Transactions on Power Systems, 28*(3), 2679–2687.
11. Li, Y., Gao, W., & Jiang, J. (2014). Stability analysis of microgrids with multiple DER units and variable loads based on MPT. In *2014 IEEE PES General Meeting| Conference & Exposition* (pp. 1–5). IEEE.
12. Yuan, H., Yuan, X., & Hu, J. (2017). Modeling of grid-connected VSCS for power system small-signal stability analysis in dc-link voltage control timescale. *IEEE Transactions on Power Systems, 32*(5), 3981–3991.
13. Amin, M., & Molinas, M. (2017). Small-signal stability assessment of power electronics based power systems: A discussion of impedance-and eigenvalue-based methods. *IEEE Transactions on Industry Applications, 53*(5), 5014–5030.
14. Rommes, J., Martins, N., & Freitas, F. D. (2009). Computing rightmost eigenvalues for small-signal stability assessment of large-scale power systems. *IEEE Transactions on Power Systems, 25*(2), 929–938.

# Chapter 9
# Cyber-Communication Network for Microgrids

## 9.1 Cyber-Communication Network

Since microgrids highly rely on the coordinated operation and control of DERs, an Information Communication Technology (ICT) infrastructure is usually utilized to transmit control signals among DERs or between DERs and MCC [1–3], which constitutes cyber-physical microgrids [4–6], as shown in Fig. 9.1.

There are several categories of the communication network, e.g., Local Area Network (LAN) [7], Metropolitan Area Network (MAN) [8], Wide Area Network (WAN) [9], Wireless [10], and Inter Network (Internet) [11], etc.

### 9.1.1 Local Area Network

Local area network is also called LAN and is usually used for the communication of private microgrid systems. It is easy to design. Several types of topologies can be considered based on the microgrid needs, e.g., star, bus, ring, tree, etc. Four typical topologies are given in Fig. 9.2.

The dispatch signals from MCC can be easily transferred over the LAN network. Since it is a private network, the data and the communication network is more secure than other communication networks. Although the LAN network is promising in securing microgrid systems, the initial setup costs of installing a LAN network is high and it can only cover a small area, leading LAN network only applicable to small-scale microgrids.

© Springer Nature Switzerland AG 2022
Y. Li, *Cyber-Physical Microgrids*, https://doi.org/10.1007/978-3-030-80724-5_9

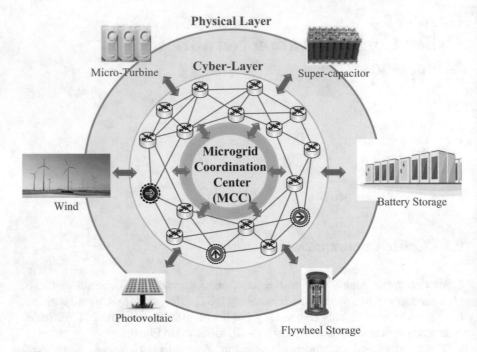

**Fig. 9.1** Illustration of cyber-physical microgrids

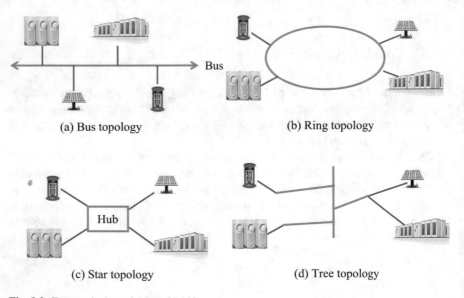

**Fig. 9.2** Four typical topologies of LAN

## 9.1.2   Metropolitan Area Network

Metropolitan area network is also called MAN. Basically, it is a bigger version of LAN and uses the similar technology as LAN. It can be used to cover a large-scale microgrid or connect multiple LANs when several microgrids need to be interconnected. MAN is efficient and can provide fast communication via high-speed carriers, e.g., fiber optic cables. But it needs more links to cover a large area. Meanwhile, it is difficult to secure the system from attacks.

## 9.1.3   Wide Area Network

Wide area network is also called WAN. WAN can be private or public network. It is usually used for a very large-scale microgrid or network microgrid systems. The communication medium used for WAN could be satellite links or public telephone networks which are connected by routers. Because WAN covers a large area, it commonly operates on low data rates. And it usually need to set up good firewalls to secure the network. Defense from hackers and viruses also introduces complexity and expense. So, security could be a potential issue when using WAN for microgrid operations.

## 9.1.4   Wireless Network

Wireless network uses the similar idea as Morse code but has better performance. We could have wireless LANs, wireless WANs, etc. For the wireless LANs, each communication host for DERs has a radio modem and antenna with which it can communicate with MCC or other hosts of DERs. It is widely used for those microgrids, where installing Ethernet is considered too much trouble or expensive.

## 9.1.5   Inter Network

Inter network or Internet is a combination of two or more communication networks. It can be formed by integrating multiple individual networks through various devices such as routers, gateways and bridges

## 9.2   Demonstration of Cyber-Physical Microgrids

Figure 8.1 is used as an example to demonstrate the outputs of DERs can be dispatched by MCC through sending control signals via the communication network. The bus topology is used to build the LAN communication network for the physical system. Two cases are carried out and introduced as follows.

*Case 1*  The microgrid is operating in the island mode. $V - f$ control is applied to control the microgrid system's frequency and voltage. At 0.2 s, MCC sends control signals to adjust the power outputs of DERs to adjust the PCC (bus 1) voltage from 1.0 p.u. to 1.01 p.u. Then at 1.0 s, MCC sends other control signals to change the PCC's voltage to 0.98 p.u. Figure 9.3 shows the changes of bus voltages in above scenarios, from which we can see bus 1's voltage can be controlled to the desired value through sending control signals from MCC and transferring those signals via the communication network. Correspondingly, the system's other bus voltages will also change.

*Case 2*  The microgrid is operating in the grid-connected mode. Table 8.3 summarizes the power outputs of DERs. At 0.2 s, MCC sends a control signal to adjust the power outputs of Micro-turbine 20. More specifically, Micro-turbine 20 will increase its active and reactive power by 100%. Then at 1.0 s, MCC sends another control signal to adjust the power outputs of Battery 31. More specifically, Battery 31 will decrease its active and reactive power by 50%. Figures 9.4 and 9.5 show the outputs of active and reactive power of DERs. Figures 9.6 and 9.7 show the changes of bus voltages and angles under the DER power changes.

From the results, we can see at 0.2 s, Micro-turbine 20's active power increases from 0.01 p.u. to 0.02 p.u. The reactive power also increases by 100%. As a result, the bus 20's voltage is significantly boosted at 0.2 s. At 1.0 s, Battery 31's active and reactive power decreases according to the control signals from MCC. As a result, the bus 31's voltage is significantly reduced at 1.0 s.

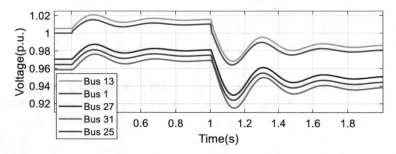

**Fig. 9.3**  Changes of bus voltages in Case 1

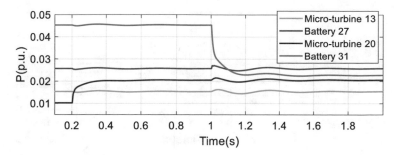

**Fig. 9.4** Changes of DER active power in Case 2

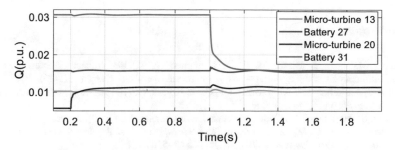

**Fig. 9.5** Changes of DER reactive power in Case 2

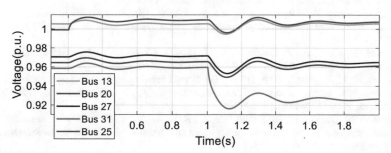

**Fig. 9.6** Changes of DER bus voltages in Case 2

## Problems

**9.1** Describe the function of the communication network in the cyber-physical microgrid systems.

**9.2** What is the typical categories of communication networks that can be applied for the microgrid system?

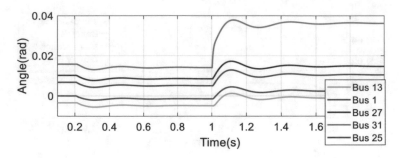

**Fig. 9.7** Changes of DER bus angles in Case 2

# References

1. Saleh, M., Esa, Y., Hariri, M. E., & Mohamed, A. (2019). Impact of information and communication technology limitations on microgrid operation. *Energies, 12*(15), 2926.
2. Dimeas, A., Tsikalakis, A., Kariniotakis, G., & Korres, 'G. (2013). Microgrids control issues. In *Microgrids* (pp. 25–80). Wiley.
3. Safdar, S., Hamdaoui, B., Cotilla-Sanchez, E., & Guizani, M. (2013). A survey on communication infrastructure for micro-grids. In *2013 9th International Wireless Communications and Mobile Computing Conference (IWCMC)* (pp. 545–550). IEEE.
4. Vu, T., Nguyen, B., Cheng, Z., Chow, M.-Y., & Zhang, B. (2019). Cyber-physical microgrids: Toward future resilient communities. Preprint. arXiv:1912.05682.
5. Deng, C., Wang, Y., Wen, C., Xu, Y., & Lin, P. (2020). Distributed resilient control for energy storage systems in cyber-physical microgrids. *IEEE Transactions on Industrial Informatics, 17*(2), 1331–1341.
6. Sridhar, S., Hahn, A., & Govindarasu, M. (2011). Cyber-physical system security for the electric power grid. *Proceedings of the IEEE, 100*(1), 210–224.
7. Fowler, H. J., & Leland, W. E. (1991). Local area network characteristics, with implications for broadband network congestion management. *IEEE Journal on Selected Areas in Communications, 9*(7), 1139–1149.
8. White, I. M., Rogge, M. S., Shrikhande, K., & Kazovsky, L. G. (2003). A summary of the hornet project: A next-generation metropolitan area network. *IEEE Journal on Selected Areas in Communications, 21*(9), 1478–1494.
9. Cahn, R. (1998). *Wide area network design: concepts and tools for optimization.* Morgan Kaufmann.
10. Cho, D.-H., Song, J.-H., Kim, M.-S., & Han, K.-J. (2005). Performance analysis of the ieee 802.16 wireless metropolitan area network. In *First International Conference on Distributed Frameworks for Multimedia Applications* (pp. 130–136). IEEE.
11. Abbate, J. (2000). *Inventing the internet.* MIT Press.

# Chapter 10
# Cyber-Physical Attacks to Microgrids

## 10.1 Cyber-Physical Attacks

Microgrids can provide flexible and reliable local green energy generation and delivery to facilitate the sustainable development of power grid. However, the distributed topology and power-electronic interfaces also offer malicious actors opportunities to access or even manipulate the bulk power grids through microgrid systems, which highly threatens the public safety and national security [1, 2]. Several power grid attacks, including the first U.S. 'denial of service' attack launched by remote hacker into the western power grid in March 2019, remind us of this issue as a severe challenge. Meanwhile, attacks on power utilities are growing in numbers now. Such disastrous attacks would not only take down a country's power grid, but could also result in catastrophic regional or national effects on public health or safety, economic security, or national security.

Recently, power system's security and resilience have been studied from several perspectives, such as reserve planning [3], event forecast [4, 5], risk assessment [6, 7], modeling and control [8, 9], cyberattacks on electricity market [10, 11], defense strategies [12, 13], recovery and repair [14, 15], etc.

Regarding the modeling of attacks to microgrids or power system, for one thing, several researches have been carried out to study the impact of cyberattack on power grids. For instance, topological models (e.g., maximum flow models [16, 17]), stochastic simulation models (e.g., Markov chain [18, 19]), statistical models (e.g., CASCADE model [20] and branching process models [21]), dynamic simulation models (e.g., ORNL-PSERC-Alaska mode models [22], COSMIC models [23] and dynamic probabilistic risk assessment model [24]), interdependent models (e.g., flocking-based hierarchical cyber-physical mode [25]), and so on.

For another, in a real microgrid system, although various types of cyber-physical attacks could occur, those attacks can be unitedly described in Fig. 10.1, which shows cyberattacks to MCC, cyberattacks to the cyber-communication network, physical attacks to the microgrids, as well as the propagation of cyber-physical attacks.

© Springer Nature Switzerland AG 2022
Y. Li, *Cyber-Physical Microgrids*, https://doi.org/10.1007/978-3-030-80724-5_10

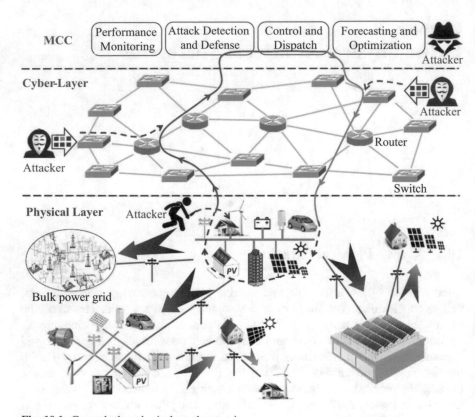

**Fig. 10.1** Generalcyber-physical attack scenario

*Cyberattacks* When the cyber-communication network is under attack, either anomalous control signals are generated due to false measurement data packets, or correct data packets get dropped, delayed, or overwritten by the malicious behavior of hackers, the physical microgrid's operation will be consequently impacted and the misbehavior could also propagate to the neighboring units or even the bulk power grids.

In microgrids, several types of cyberattacks could happen [26–29], e.g., malware attack to MCC, man-in-the-middle attack to steal measurement data, denial-of-service attack, internet of things attacks, etc.

*Physical Attacks* When the physical layer is under attack and MCC is unable to detect and defend those attacks, the microgrid's operation can be directly impacted and aggravated, because the microgrid's operation highly relays on the control signals from MCC, which are generated based on the measurements of the compromised system. So, it becomes a cascading process. The influence can quickly propagate to the neighboring units.

## 10.2  Impact of Communication Network

The operations of microgrids highly depends on the timely information exchanges between DERs and MCC through the communication network. So, the communication network is playing a very important role. Since microgrids have low inertia due to the adoption of power-electronic interfaces when integrating DERs, microgrids become sensitive to cyber or physical events, such as the delay of sending data packets caused by the communication network latency.

To demonstrate the impact of communication network latency on the physical microgrids' operations, Fig. 8.1 is also used as an example. At 0.2 s, Load 14's active power increases by 20%. The baseline communication latency is 0.01 s. Five latency cases are carried out. In case 1, 10% increment of communication latency is considered, i.e., 0.011 s. For case 2, it is 100% latency increment, i.e., 0.02 s. For case 3, it is 200% latency increment, i.e., 0.03 s. For case 4, it is 210% latency increment, i.e., 0.031 s. For case 5, it is 230% latency increment, i.e., 0.033 s.

The bus 13's voltage in the above five cases are shown in Fig. 10.2. Eigenvalues at 0.0 s in the above five cases are shown in Fig. 10.3. From Fig. 10.2, we can see in the same condition of the physical system, as the communication network latency

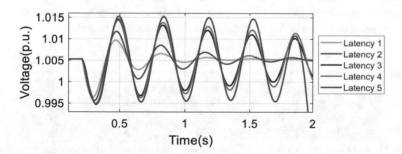

**Fig. 10.2** Voltages under different communication latency

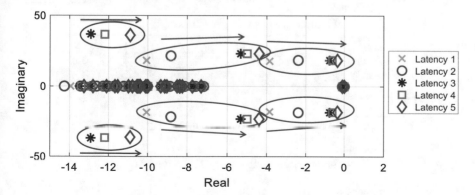

**Fig. 10.3** Eigenvalues under different communication latency

increases, it leads to more dynamics, i.e., the physical microgrid system becomes less stable. The eigenvalue distributions in Fig. 10.3 also show critical eigenvalues are moving to the right hand as the communication network latency increases, which verifies the system becomes less stable.

## 10.3   Demonstration of Cyber-Attack

The operations of microgrids also highly depends on the secure communication network which is used to sending the measurement data packets to MCC and dispatch signals to DERs. So, the security of the communication network is playing a very important role.

To demonstrate the impact of cyberattack to communication network on the physical microgrids' operations, Fig. 8.1 is also used as an example. The system is operating under droop control, i.e., the island mode. The attacker eavesdrops the communication between MCC and Battery 31. At 1.5 s, the attacker overwrites the power dispatch signals of Battery 31 and injects the wrong signals to the communication network for maliciously changing its power outputs. Figure 10.4 shows the instantaneous voltage and current responses of Battery 31. Figure 10.5 shows the instantaneous voltage and current responses of Micro-turbine 20. Figure 10.6 shows the frequency response of the system under the above cyberattack scenario.

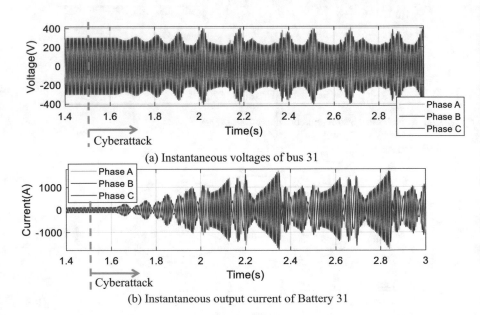

(a) Instantaneous voltages of bus 31

(b) Instantaneous output current of Battery 31

**Fig. 10.4**  Instantaneous voltage and current responses of Battery 31 under cyberattack

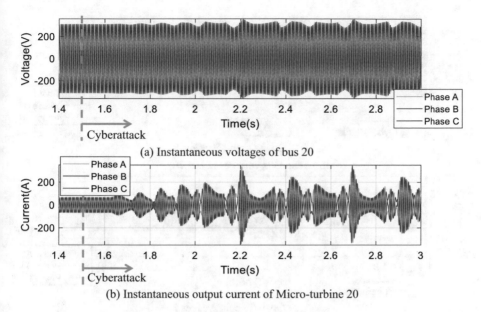

(a) Instantaneous voltages of bus 20

(b) Instantaneous output current of Micro-turbine 20

**Fig. 10.5** Instantaneous voltage and current responses of Micro-turbine 20 under cyberattack

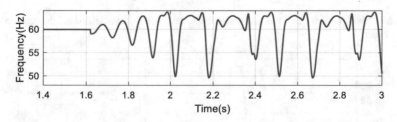

**Fig. 10.6** Frequency response of the system under cyberattack

From Fig. 10.4, we can see the voltage and current of Battery 31 are significantly influenced because of cyberattack. Meanwhile, the compromised Battery 31 can also impact other DERs as shown in Fig. 10.5. Figure 10.6 also demonstrates cyberattack can significantly impacts the island microgrid system's frequency.

Because of the severe influence of cyberattacks, based on system monitoring and detection, if cyberattacks can be quickly detected to isolate the compromised DERs, the impact of cyberattacks can be mitigated. In the above scenario, Circuit Breaker 5 is open to disconnect the compromised Battery 31. Figure 10.7 demonstrates the Micro-turbine 20's voltage and current responses. The frequency response is shown in Fig. 10.8, from which we can see after isolating the compromised Battery 31, the remaining island microgrid system can be controlled to an equilibrium point, so that the frequency can be stabilized.

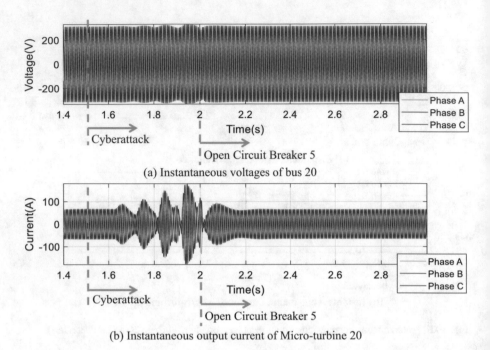

(a) Instantaneous voltages of bus 20

(b) Instantaneous output current of Micro-turbine 20

**Fig. 10.7** Micro-turbine 20's voltage and current responses when defending the system from cyberattacks

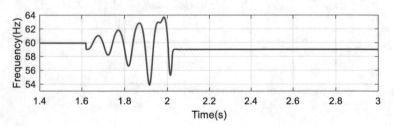

**Fig. 10.8** System frequency responses when defending the system from cyberattacks

## 10.4   Demonstration of Physical Attack

It is straightforward that the physical attacks to DERs or microgrids will significantly influence the operations of the systems. To demonstrate the impact of physical attacks, Fig. 8.1 is also used here as an example. The system is still operating under droop control, i.e., the island mode. At 1.5 s, the inverter's controller of Micro-turbine 13 is under attack, causing it not to able to control Micro-turbine 13. More specifically, the inverter's controller is disconnected, so the control signals received by the inverter becomes zero. The inverter controllers of PV 18, Battery 27, and Battery 31 are under the same attack at 1.6, 1.7, and 1.8 s, respectively. Figure 10.9

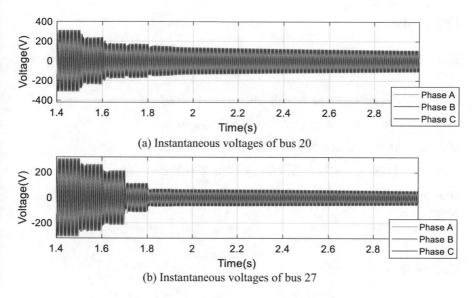

Fig. 10.9 Instantaneous voltage responses of bus 20 and bus 27 under physical attack

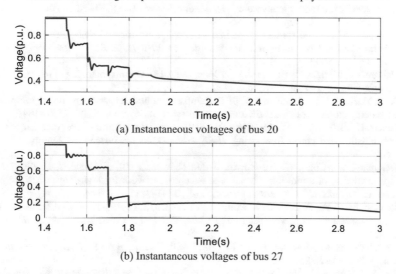

Fig. 10.10 RMS values of bus voltage at node 20 and node 27 under physical attack

shows the instantaneous voltage responses of bus 20 and bus 27. Figure 10.10 shows the RMS values of bus voltage at node 20 and node 27. From the results, we can see the physical attacks to DERs significantly influence the operations of the systems, as the voltage amplitudes are severely reduced.

## Problems

**10.1** Please describe the function of three subsystems of the conventional power grid.

**10.2** Why is it necessary to develop microgrid systems?

**10.3** Please describe a microgrid system and give an example.

**10.4** What is the function of the cyber layer of CPMs?

**10.5** What is the typical control strategy of photovoltiac system?

**10.6** Please describe the bi-directional power flow phenomenon and explain what causes it.

## References

1. Stamp, J. E., Veitch, C. K., Henry, J. M., Hart, D. H., & Richardson, B. (2015). Microgrid cyber security reference architecture (v2). Sandia National Lab.(SNL-NM), Albuquerque, NM (United States), Tech. Rep.
2. Stamp, J. (2012). The spiders project-smart power infrastructure demonstration for energy reliability and security at us military facilities. In *2012 IEEE PES Innovative Smart Grid Technologies (ISGT)* (p. 1). IEEE.
3. Quashie, M., Marnay, C., Bouffard, F., & Joós, G. (2018). Optimal planning of microgrid power and operating reserve capacity. *Applied Energy, 210*, 1229–1236.
4. Gong, Y., Jiang, Q., & Baldick, R. (2015). Ramp event forecast based wind power ramp control with energy storage system. *IEEE Transactions on Power Systems, 31*(3), 1831–1844.
5. Gilbert, C., Browell, J., & McMillan, D. (2019). Leveraging turbine-level data for improved probabilistic wind power forecasting. *IEEE Transactions on Sustainable Energy, 11*, 1152–1160.
6. Veeramany, A., Coles, G. A., Unwin, S. D., Nguyen, T. B., & Dagle, J. E. (2017). Trial implementation of a multihazard risk assessment framework for high-impact low-frequency power grid events. *IEEE Systems Journal, 12*(4), 3807–3815.
7. Bhuiyan, M. Z. A., Anders, G. J., Philhower, J., & Du, S. (2019). Review of static risk-based security assessment in power system. *IET Cyber-Physical Systems: Theory & Applications, 4*, 233–239.
8. Tabatabaei, N. M., Kabalci, E., & Bizon, N. (2019). *Microgrid architectures, control and protection methods.* Springer.
9. Han, Y., Zhang, K., Li, H., Coelho, E. A. A., & Guerrero, J. M. (2017). MAS-based distributed coordinated control and optimization in microgrid and microgrid clusters: A comprehensive overview. *IEEE Transactions on Power Electronics, 33*(8), 6488–6508.
10. Wang, B., Dabbaghjamanesh, M., Kavousi-Fard, A., & Mehraeen, S. (2019). Cybersecurity enhancement of power trading within the networked microgrids based on blockchain and directed acyclic graph approach. *IEEE Transactions on Industry Applications, 55*(6), 7300–7309.
11. Poyrazoglu, G., & Oh, H. (2018). Impact of cyber-security breach to price signals on power market: an experimental human simulation. *Mediterranean Conference on Power Generation, Transmission, Distribution and Energy Conversion (MEDPOWER 2018)*, pp. 1–7.

12. Chen, Y., Hong, J., & Liu, C.-C. (2016). Modeling of intrusion and defense for assessment of cyber security at power substations. *IEEE Transactions on Smart Grid, 9*(4), 2541–2552.
13. Ashok, A., Govindarasu, M., & Wang, J. (2017). Cyber-physical attack-resilient wide-area monitoring, protection, and control for the power grid. *Proceedings of the IEEE, 105*(7), 1389–1407.
14. Tabatabaei, N. M., Ravadanegh, S. N., & Bizon, N. (2018). *Power Systems Resilience: Modeling, Analysis and Practice*. Springer.
15. Golshani, A., Sun, W., & Sun, K. (2017). Advanced power system partitioning method for fast and reliable restoration: toward a self-healing power grid. *IET Generation, Transmission & Distribution, 12*(1), 42–52.
16. Wenli, F., Ping, H., & Zhigang, L. (2016). Multi-attribute node importance evaluation method based on gini-coefficient in complex power grids. *IET Generation, Transmission & Distribution, 10*(9), 2027–2034.
17. Fan, W., Huang, S., & Mei, S. (2016). Invulnerability of power grids based on maximum flow theory. *Physica A: Statistical Mechanics and its Applications, 462*, 977–985.
18. Zhan, C., Chi, K. T., & Small, M. (2016). A general stochastic model for studying time evolution of transition networks. *Physica A: Statistical Mechanics and its Applications, 464*, 198–210.
19. Hines, P. D., Dobson, I., & Rezaei, P. (2016). Cascading power outages propagate locally in an influence graph that is not the actual grid topology. *IEEE Transactions on Power Systems, 32*(2), 958–967.
20. Wu, H., & Dobson, I. (2013). Analysis of induction motor cascading stall in a simple system based on the cascade model. *IEEE Transactions on Power Systems, 28*(3), 3184–3193.
21. Dey, P., Mehra, R., Kazi, F., Wagh, S., & Singh, N. M. (2016). Impact of topology on the propagation of cascading failure in power grid. *IEEE Transactions on Smart Grid, 7*(4), 1970–1978.
22. Carreras, B. A., Newman, D. E., Dobson, I., & Degala, N. S. (2013). Validating OPA with WECC data. In *2013 46th Hawaii International Conference on System Sciences* (pp. 2197–2204). IEEE.
23. Song, J., Cotilla-Sanchez, E., Ghanavati, G., & Hines, P. D. (2015). Dynamic modeling of cascading failure in power systems. *IEEE Transactions on Power Systems, 31*(3), 2085–2095.
24. Henneaux, P., Labeau, P.-E., Maun, J.-C., & Haarla, L. (2015). A two-level probabilistic risk assessment of cascading outages. *IEEE Transactions on Power Systems, 31*(3), 2393–2403.
25. Wei, J., Kundur, D., Zourntos, T., & Butler-Purry, K. (2012). A flocking-based dynamical systems paradigm for smart power system analysis. In *2012 IEEE Power and Energy Society General Meeting* (pp. 1–8). IEEE.
26. Liu, X., Shahidehpour, M., Cao, Y., Wu, L., Wei, W., & Liu, X. (2016). Microgrid risk analysis considering the impact of cyber attacks on solar PV and ESS control systems. *IEEE Transactions on Smart Grid, 8*(3), 1330–1339.
27. Chlela, M., Joos, G., & Kassouf, M. (2016). Impact of cyber-attacks on islanded microgrid operation. In *2016 Proceedings of the Workshop on Communications, Computation and Control for Resilient Smart Energy Systems* (pp. 1–5).
28. Li, Y., Qin, Y., Zhang, P., & Herzberg, A. (2018). SDN-enabled cyber-physical security in networked microgrids. *IEEE Transactions on Sustainable Energy, 10*(3), 1613–1622.
29. Li, Y., & Du, L. (2021). Programmable and reconfigurable cyber-physical networked microgrids through software-defined networking. in *2021 IEEE Transportation Electrification Conference & Expo (ITEC)* (pp. 1–5). IEEE.

# Index

**B**
Battery, 41

**C**
Cyberattack, 188

**D**
Differential-algebraic equations (DAEs), 165
Doubly-Fed induction generator (DFIG), 91
Droop control, 153

**E**
Eigenvalue, 174

**F**
Flywheel, 43

**I**
induced voltage, 61
Inverter, 99
'I-U' curves, 24

**L**
Latency, 187

**M**
Main rotating magnetic flux, 58
Maximum Power Point Tracking (MPPT), 28
Microturbine, 55
Modulation index, 103

**P**
Photoelectric Effect, 13
Photons, 14
Physical attack, 190
Power factors, 68
Power outputs, 127
'P-U' curves, 25

**S**
Slip, 83
State of Charge (SoC), 41
Supercapacitor, 50

**T**
Tip speed ratio, 78

**V**
Vector control, 144

**W**
Wind energy, 77

© Springer Nature Switzerland AG 2022
Y. Li, *Cyber-Physical Microgrids*, https://doi.org/10.1007/978-3-030-80724-5

Printed in the United States
by Baker & Taylor Publisher Services